Beekeeping
Mentor in a Book

by Donald P. Studinski

Published by:
X-Star Publishing Company, founded 1961
xstarpublishing.com

To Contact the Author, Donald P. Studinski:
by email: dstudin@yahoo.com
by postal service: POB 1995, Broomfield, CO 80038
author's website: www.HoneybeeKeep.com

You can join Don's Yahoo Group. It's free. Read lessons and ask questions. Share
experiences with other students.

https://groups.yahoo.com/neo/groups/BeekeepingStudents/info

Cover Photo Credit: Betsy Seeton

Chapter Photos Credit: Steve Kennedy

"Whenever a Taker couple talk about how wonderful it would be to have a big family, they're reenacting this scene beside the Tree of Knowledge of Good and Evil.
They're saying to themselves,
'Of course it's our right to apportion life on this planet as we please. Why stop at four kids or six? We can have fifteen if we like. All we have to do is plow under another few hundred acres of rain forest— and who cares if a dozen other species disappear as a result?'"

… <u>Ishmael</u> by Daniel Quinn, p 181, Bantam trade paperback edition / June 1995

"We have discussed food, nonrenewable resources, and pollution absorption as separate factors necessary for the growth and maintenance of population and industry. We have looked at the rate of growth in the demand for each of these factors and at the possible upper limits to the supply. By making simple extrapolations of the demand growth curves, we have attempted to estimate, roughly, how much longer growth of each of these factors might continue at its present rate of increase. Our conclusion from these extrapolations is one that many perceptive people have already realized—that the short doubling times of many of man's activities, combined with the immense quantities being doubled, will bring us close to the limits to growth of these activities surprisingly soon."

… <u>The Limits to Growth</u> by Donella H. Meadows, Dennis L. Meadows, Jorgen Randers and William W. Behrens III, p 88, Potomac Associates / December 1973

This book is dedicated to my parents:
Rush B. Studinski
Nancy L. Studinski
who have tolerated much both giving me life and keeping me alive these many years.

And to:

Marcia L. Heiser
who quietly holds me up for the world to see.

Table of Contents

Illustration Index

Forward

© steve kennedy 2014

Find here the story of a beekeeping mentor and his students just as it occurred in 2012. This story is offered to the reader as a way of experiencing the relationship between a mentor and a student without having been present. Words highlighted in **bold** are potentially new vocabulary. You can find these defined in the glossary. Readers can learn from real experiences that happened to the author and his students during 2012 which provide examples about the **beekeeper**'s life. Laced with general honeybee and beekeeping knowledge, as well as valuable topical discussion, you will find entertainment as well as reference material. Imagine yourself as a student having dialog with your mentor and your fellow students. Use the pictures to imagine yourself at the **apiary** seeing and doing what you are studying. The reader can learn what a beekeeper needs to be thinking and doing in any specific time of year. This text covers mentoring that happened in Colorado, USA. Thus, you must use imagination to adjust this discussion to your own climate and seasonal changes.

For any given month of activities, extensive preparation and planning has been in progress, sometimes for several months, in advance. Let's say you want to learn about **split**s and you see that it's covered in the chapter about April. You should also review the information in the several months leading up to April because that's when all the planning happened that leads up to April splits.

*Illustration 1: Come together in the spirit of community and build seven **Warré** hives in a day. Photo credit: Living Systems Institute (LSI)*

I am in the habit of naming my colonies. This aids in our student discussions. Over time, students learn where each **colony** is and we can talk about them specifically by name. This text also refers to colonies by name. This will not cause confusion, but rather will allow the reader to follow a colony through the calendar year as they progress. This enhances the readers experience.

This is really two books in one. Every second chapter is the story of beekeeping in 2012, labeled by month. Student names are fictional, not intended to refer to or imply any real living person, however, the events are real and real students participated as indicated. This story is interrupted every second chapter to provide reference material not specifically related to the 2012 beekeeping season. Chapters are intentionally short with plenty of photographs. They are meant to be read in a single engagement with the book. Then, the reader can walk away and let the material sink in over time. My suggestion for the book is that you read it at bedtime the night before you plan an apiary visit. Read that month's chapter, sleep on what you have read, and you will be ready in the morning for your apiary visit. An extensive index is included for when you need to find information on a specific topic. So it is that this text is intended to serve the **bees**, and their human beneficiaries, arranged monthly for handy reference. You can follow along as if your mentor was right with you!

January ... Always Think Ahead

Congratulations, you're a **beekeeper**! Living **bees**, by the thousands, are now counting on you.

When you arrive at the apiary to get some work done, if you have not planned ahead, you may find yourself without a tool or a piece of equipment you need. It's disappointing to have to make another trip. With bees, you want to be prepared. Planning a year in advance is not too far ahead for some of the manipulations you are going to want to do. At a minimum, you need to have planned several months ahead. With time, you will see the need for this.

It's January 1, 2012 and we have four living colonies of honeybees. We're making plans for how these bees will be used in the coming year. Some of the plans will become our reality. Others will not. That's one of the biggest parts of keeping bees: making plans and adjusting to reality. We think these four colonies will become seven through the splitting process. Nice plan. Let's see how close it comes to reality.

If you are new to beekeeping and you want to have bees in the coming year, you should have your bees ordered by now. Bees are increasingly scarce and increasingly expensive. The vendor has to plan months in advance to get your bees ready. You should expect to pay in advance and don't be surprised if there are delays in delivery by days or weeks. This is nature. Try as we might, the end result is vulnerable to unexpected weather events.

Cold weather can also be an ideal time to ready your **hive** equipment. Repairs may be necessary. You might assemble new equipment. Get your hive in position and leveled in anticipation of the new colony to come.

Expect the Unexpected

News came 12/31/2011 that the colony in the **top-bar hive (TBH)** at Living Systems Institute (LSI), the Longmont girls, was overturned by **wind**. This is, at least, a significant set back, and at most, a death sentence for this colony. It's all part of being "in" nature. If they were chilled or if the **queen** is hurt or dead, then they are doomed. This is a perfect example of how our beekeeping learning opportunities may come suddenly and will frequently be weather related. I'm heading to LSI now, and plan to spend a good bit of the afternoon there. I suspect I will be there tomorrow also, the temperature tomorrow is supposed to be in the 50s, which is key. Above 50°F, bees can move about freely and a beekeeper can open the hive without killing them with cold. I have not completed my plan of action yet. I'm

Illustration 2: Top-bar hive. This is the hive that blew over. It's nailed to those stumps! The whole thing went over. Photo credit: David Braden

considering moving the colony into a **Langstroth** hive. I have to plan out the equipment and the place.

More planning.

We anticipate capturing four **swarms** this year. There is no way to plan for this better than just knowing that swarm season starts in April and will run through June. During this time, students may get a call that a swarm is available at a certain address. They are invited to roll on that call. If they can participate, then they will learn a lot. On the other hand, if the timing doesn't work, that's not a problem, there will be other opportunities. This would bring us to 11 colonies, assuming all goes as planned.

Illustration 3: Langstroth hive, five frame nuc, migratory cover, honeybee package. Photo Credit: Don Studinski

Unfortunately (or fortunately, depending on how you look at it), 11 colonies is not enough to fill all the hives we have available. Luckily, we have students willing to purchase bees. Although I prefer not to purchase bees, there are times when it may be necessary. I prefer to get local stock whenever possible. But, finding bees for sale in Colorado is not easily done. My first choice in advising about your bee purchase is to point you to Minnesota Hygienic bees. This strain of Italian bees was developed from the research of Dr. Marla Spivak at the University of Minnesota. I have friends who bought these bees and bragged about their honey production. That is the basis of my recommendation. I have not purchased bees by mail, but if I was buying, this is what I would choose. These are expensive.

Here's a little story about ordering bees:

In January, 2010, I knew I would need additional colonies, so I ordered two **nucs** (nucleus colonies) from New Mexico. The advantage of these bees is they are bred in high elevation, so they would be better acclimatized to our local Colorado weather conditions than a colony from somewhere like Texas. These were $150 each and came in a cardboard nuc box, five **frame**s. My friend, Eric, picked up our bees personally in New Mexico and drove them all the way to Denver the same day. Unfortunately, one of my nucs got overheated in transit (not the vendor's fault). The bees never thrived and had to be combined with another colony before the first summer was done. The second colony seemed to do well for a first-year colony. They did not produce surplus honey, but that was not a surprise. The bee population seemed strong. They survived the winter of 2010-2011 and I had high hopes for a honey harvest in 2011. Well, that didn't happen either. They seemed healthy enough, but produced nothing for harvest. I ended up trying to force a **supersedure** (replace the queen). This resulted in them swarming to the point that they had too small a population to defend the hive. In the end, they were **robbed** to death. My point is: this was a $300 attempt that produced nothing in 2 years of effort. You too will experience failures along the way. You will also experience joys. Be prepared for both and give it time.

Dealing with Disaster

Thankfully, Colorado provided us with the ideal day, January 2, to work the bees in winter soon after the big wind that turned over the Longmont girl's top-bar hive and brought down one of the big trees at LSI. Ruth and David were there to help. These bees had to brave two cold nights in the teens (degrees Fahrenheit) in that jumbled mess of broken **comb** and spilled honey. But it looks like they came through fine.

We arrived just after 11AM while it was still too cold to open up the hive. This gave us plenty of time to get the new Langstroth hive set up, ready to accept the girls. We used one **deep** for the new hive setup with four frames of honey as the outside frames and five frames of empty drawn comb in the middle (nine frames in a ten frame deep; extra space left on the outside). It was probably around noon that we started seeing bees in the air. This is the perfect indication that the temperature is now ready for us to open the damaged top-bar hive. Once we opened it up, it became clear that leaving them in place (which was the plan B option) would not be the best plan. The seven frames of **brood** were a broken mess; four broken off the top bars, three still in tact. Her highness is a very

Illustration 4: Violet colony in killer snow and cold. This wiped out all the blossoms. Photo credit: Don Studinski

productive queen as demonstrated by her **capped brood** and **larva** approximately the size of two golf balls. I had expected no brood at all. I am fairly sure she is alive and well in the Langstroth hive because I saw the bees migrating from the top-bar disaster to the Langstroth with purpose. This behavior indicates there is a live queen in the destination. Here's why: If the queen was still in the top-bar mess, the bees would have wanted to stay there. Further, the bees we had dropped into the

Langstroth would have been moving back to the top-bar. On the other hand, if there was no queen alive, then the disoriented bees would have shown no preference for either the top-bar (old home) or the Langstroth (new home). Whoever found themselves in the top-bar would have just stayed there and likewise in the Langstroth. There would be no strong pull one way or the other.

This was a sticky, messy job. We were covered with honey on our hands the whole time. By the time we completed the move, about 1PM, most of the bees were settling in to the new home with clear preference for that over the top-bar. This is a clear indication that a live queen is within the new home. We had a quick lunch and returned to clean up the mess of comb and honey left behind in the top-bar hive. This took until near 3PM, but leaves us with a top-bar hive covered in honey (inside and out) and basically ready to accept a new colony which will have to start over **drawing out new comb**.

It's not every January 2nd we get to work the bees because of our Colorado winters. Ruth got only one sting, because she had not cuffed her pants and a bee went up from ankle level. I got three stings, one like Ruth's, one on my neck & one on my hand.

What Should be Ready Now

If you are going to have a hive on your property this year, your hive stand should be in place now. Since we live in the northern hemisphere, you will want the front entrance of your hive to face south (for full **sun** exposure) and the stand should provide a slight tip downward toward the front of the hive (to ensure rain runs out). My stands are built with two or three cinder bricks on the ground and two landscaping timbers placed on the cinder bricks. I can easily put two or three hives on a stand like this. Be sure your chosen hive placement accounts for a wind break on the north and west sides (we get prevailing winds from those directions). Notice the hay bales in the picture. Those are acting as wind break in this case. Don't forget about access to **water**. Your bees must have water. They will find the most convenient source, even if it's your neighbors drippy faucet.

Quite a few things are of note in that picture. To the right of the tall Langstroth is a wooden nuc with a circular entrance reducer This is a deluxe model as nucs go. It's very handy for moving a colony. The circular piece over the entrance lets you set it to allow various sized bees to pass through or not. There is a setting to allow no bees to pass which I use to transport bees in my car. Other settings let only **worker**s pass, workers and **drone**s or all bees (wide open). It's a six frame nuc. Nucs also come in three, four or five frame size. Further to the right is a migratory outer cover; different from a **telescoping cover**. Notice how it would allow two hives to sit next to each other without a gap of space between. That's the advantage. The last thing to the right is an old **bee package**. This shows you an example of how bees get shipped. The circular cut in the top is where the syrup feeding can rests. It's also where you dump the bees out. Notice the carpet in front of the hive stand. This suppresses the grass leaving the airport clear and suggests a safe distance to stand clear for humans. Lastly, notice various wind-breaking materials in the background. You will want to take note of your environment when selecting an appropriate hive placement.

You should have picked out your **bee suit** and ordered it by now. At a minimum, you need a **veil**. I recommend that you have gloves and some way to seal off your pant legs at the ankles. Consider wearing boots to cover those ankles or expect that bees will sting through your socks. It is not realistic to think you can be a beekeeper and not get stung. This means that, even if your chosen suit is a full body suit, you should expect to get some stings. Do you know if you are allergic to bee **venom**? If not, do not inspect bees without a partner until you have been stung and you know how you react. Consider taking an intentional sting. This can be done in a controlled manner with easy access to medical facilities as appropriate. Consider taking a second intentional sting after the first one has run its course. Some say the severe allergic reaction doesn't show itself on the first sting.

Now, let's get familiar with some hive vocabulary. Starting at the bottom and working our way up, a Langstroth hive is made of these parts:

bottom board – a plastic, wooden or screened "board" on which the first hive body box will rest.
deep - a box intended for holding frames. Deeps are usually 9 5/8 inches deep for a standard Langstroth box. This is the area for the queen to raise brood. It is generally not used for honey.
deep - a second brood box. We want a lot of brood.
queen excluder - a plastic or wire grid within a plastic or wood frame. Spaces between are big enough for worker bees to pass, but too small for drones or a queen to pass. The purpose of this is to keep **egg** laying below the queen excluder in the deeps. We don't want eggs or larva in our honey.
super - a smaller box intended for holding frames. Medium supers are 6 5/8 inches deep for a standard Langstroth box. This is the area where honey is stored. This term is also used for ANY box that is placed <u>above</u> the brood area no matter what the dimensions are, so this can be a point of confusion. For example, if I wanted to collect honey in a deep and I put it over the brood area, then I would refer to it as a super. Since honey is heavy, it's usually more convenient to store in a smaller super.
inner cover - a flat board, sometimes framed, with a hole in the center which allows bees to pass through. This is analogous to the ceiling in a home. It's important for when you want to use a top feeder. It also provides some insulation.
telescoping cover - a wooden roof, sometimes covered with metal, which not only covers the top, but comes down a bit on each side of the hive (this ensures it doesn't fly away easily from wind) and prevents rain from entering the hive. Commercial guys don't like the overlapping part because it takes up extra room, so you will see some covers that are not telescoping.

If you already have hives with living bees in the apiary, now is an excellent time to make repairs to any damaged equipment. Some might need an external coat of paint (see February for more detail). You might also want to move equipment around. Where do you need another empty hive, in anticipation of spring? Where do you need your supers? You can get some of those chores done inside during the cold days of January.

Illustration 5: First bloom 20130404, photo credit: Bobbi Storrs

Making Plans

We will target 3/25, 26, 27 to start our splits; just after the spring equinox. If that is rained or snowed out, we will use 4/1, 2, 3. We can do this because I know that, in 2011, I had **capped drone cells** on 3/12. This means drones will have **emerged** no later than 3/25 (13 days after being capped). Once you have male bees available, you can make new queens and get them serviced. Queens take 16 days, so, theoretically, you could start them 16 days before 3/25, or 3/9, but that's too early in our climate and I wouldn't want to cut it that close; those drones must be mature and plentiful when the virgin queens fly for mating. Assuming we successfully split on 3/25, we will have new **capped queen** cells by 4/1 and queens emerging on 4/8 (Easter). Give her one week to mature and a second week to fly for mating. At that point, she should be home and laying eggs, that's 4/22 (Earth Day). My visual impairment keeps me from seeing eggs, so I wait another week before inspecting. I want to be able to see larvae and possibly some capped brood 4/29. That's when we know the split has been successful. Two weeks later, 5/13 (Mother's Day), new honeybees are emerging and the population of the colony starts growing quickly. The first tree bloom opens about 4/9, so, the **nectar flow** is on during this time which keeps the bees fed and healthy.

What if my split is not successful? What if I don't see capped brood 4/29/2012? The answer is that I need to be prepared to purchase a new fertile queen on short notice. Alternatively, I could **merge** sisters back together; a forfeit of the split entirely. The split "family" goes right back with the mother colony; they are all sisters anyway.

Here's what you need to know to calculate a plan like this. You might want to memorize this information. The day the egg is laid is day one:
Queen: capped on day 9, emerge on day 16
Worker: capped on day 9, emerge on day 21
Drone: capped on day 11, emerge on day 24

Illustration 6: Warré Installed on four leg stand. Photo Credit: David Braden

The first swarm of 2011 was seen in Longmont on 4/25/2011. We are making a plan that closely mirrors what the bees are doing anyway. If they can swarm naturally 4/25, then they are clearly ready to handle virgin queens in that time frame. We can back off that date to see how to lay out our split plans. One of the **permaculture** principles we want to practice is *observe and interact*. When we plan our split efforts around what we see nature doing, we are practicing that principle. We also get our nucs on 4/21/2012. Very similar timing. A busy time of year with a lot of new bees.

Thinking of Spring

In 2011, the **first blooms** were seen 4/9/2011. You need to make a habit of marking these things in your calendar because it's very helpful to be able to look back and see what happened a year ago, two years ago, etc. I also mark on my calendar every time I see live bees flying around the entrance over winter. That way, I can pretty accurately identify when death occurred, if I run into a **dead-out**, and I can connect the death to weather, starvation or disease.

Let's suppose you are a colony of bees and you want to hit the **flow**, mid April, with 50,000 bees. You are gearing up for that now, in January. Your queen can lay, conservatively, 1000 eggs each day, but not this early in the year. She cannot lay more eggs than can be kept warm by the current **cluster**. Because your bee population is at its smallest right now, her brood nest is very limited, maybe 100 bees, 50 on each side of a frame. That's approximately the size of a golf ball diameter. Maybe there are enough bees to do two frames like this. It all depends on the population of workers left. Each brood cycle takes 21 days (workers). Cycles of brood overlap in time. Each time they emerge, your colony immediately starts another cycle, this time a bit bigger, because you have more workers. By mid February, we should be able to see a noticeably larger population at the front door.

About that time, the girls will begin preparation for swarm season. The first order of business is to produce some males. There are no males in the colony in January, we will explain why when we get to fall. But soon, males will be worth having around because the bees will be preparing for virgin queens. Around here, the bees' target date for an emerging drones will be mid to late March. The eggs have to be laid 24 days ahead of that. Roughly, the end of February. Drones get capped after 11 days. Therefore, when we are doing **first inspection**, about mid March, we are hoping to see capped drones. When we start seeing the capped drones, we can start anticipating swarm season, make final preparations for splits and start swarm prevention efforts.

This year we will perform first inspection 3/11/2012, weather permitting. Backup dates are 3/12, 3/13 or 3/18. We will begin with the TBH at the Broomfield Organic Garden. We will work our way south from there hitting each hive along the way. We will be using the **smoker**. Expect to spend some time getting it lit each stop. That night we will move hives in anticipation of splits. We need some free space next to hives that need to be split.

One last reminder: beekeepers are very busy in April running swarm calls, installing honeybee packages, making splits, checking queen health and more. Now that you are a beekeeper, it's probably a good idea to be in the habit of getting your taxes done early. It can be very inconvenient to have to miss beekeeping in order to do something really "fun" like taxes.

Welcome to beekeeping. Think about your colonies. Anticipate their needs. Make a plan. Then, adjust.

Facts You Should Know

© steve kennedy 2014

Here are some basic facts new beekeepers should learn about honeybees. The public is fascinated by honeybees and you will get peppered with questions as soon as anyone finds out you are a beekeeper. The more of this you can remember off the top of your head, the more they will enjoy listening to you talk about your bees and beekeeping. That can translate into honey sales.

Queens, Workers and Drones

Honeybees develop from an egg (embryo) to an adult (imago) through four stages: egg, larva, **pupa** and **adult**. This is called **metamorphosis** meaning that the insect's body is completely transformed. Complete honeybee metamorphosis is called **holometabolism**. Insects that go through holometabolism are called holometabolous insects.

It's common knowledge that honey bees have a **queen** in the colony. The queen is always referred to in the singular because there is, almost always, only one per colony. Rare exceptions of multiple queens do exist, but those are not covered here. The other two types of bees in the colony are the drones (males) and the workers (females). We call the females "workers" because they literally do ALL the work. We'll learn the chores later. In this image of the three castes, the queen is to the left, the worker is in the middle and the drone is to the right.

Illustration 7:Worker, Drone, Queen. Photo credit: Zachary Huang

The queen is the only female within the colony with a fully developed reproductive system. Only the queen can mate. Therefore, only the queen can lay a **diploid** (fertilized by a sperm) egg which will become a female bee. Queens also lay **haploid** eggs which become drones, a male bee. Workers guide her to a cell in which they want her to lay. She measures the cell width with her front legs[1] and, based on the size of the cell, she lays either a fertilized egg (worker) or an unfertilized egg (drone).

Workers sometimes do lay haploid (unfertilized) eggs. This is most commonly seen by beekeepers when the colony becomes queenless. This is called a **laying worker**. Workers do not produce nicely organized brood patterns like a healthy queen would. Rather, they produce many bullet shaped drone cells throughout the brood area, usually badly scattered. This is a bad sign that the colony has been queenless for quite some time and they are probably doomed to die. Immediate queen replacement is required and the laying worker must be eliminated. This can be quite challenging as we will learn later.

An egg is an egg for three days, then it **hatch**es into a larva. During the first three days as a larva, all honeybees, male and female, queen and worker, receive the same feeding of **royal jelly**[2]. Therefore,

---

1   http://www.extension.org/pages/21744/worker-queen-differentiation-basic-bee-biology-for-beekeepers, visited 7/30/2014

2   http://en.wikipedia.org/wiki/Royal_jelly, visited 1/28/2014and Beekeeping for Dummies, Wiley Publishing, 2002, pg 33

every female larva is a candidate to become a queen up to and including day six. Up to that point, the larvae get fed royal jelly an average of 1300 times each day. At that point, worker and drone diet changes over to honey and bee bread.

If this larva is intended to be a queen, the royal jelly feedings continue. The diet fed during days seven, eight and nine are what makes a queen. The queen bee develops into an adult in 16 days: three days as an egg, six days as a larva and seven days as a pupa. The **queen cell** gets capped on the 9th day. The queen emerges on the 16th day. After emerging, the virgin needs about a week to sexually mature and then a second week to take her **nuptial flights**, where she will mate with 10 to 20 drones. At that point, she will settle into her home and begin to lay. She can lay as many as 2000 eggs per day. She will specifically lay fertile eggs in worker cells and infertile eggs in drone cells. Her highness' attendants will see to her every need, including feeding and waste removal, and especially heating and cooling. She will continue to be fed royal jelly for her entire life.

That tiny window of about one week for her to successfully mate is the only time in her life that she can have sex. If the weather prevents successful mating flights, she will not be capable of performing her duties. A healthy queen can live for up to nine years, but we don't see that happening these days.

Failing queens are sometimes killed by the workers intentionally. If they have a young larva, then they can make a new queen. But, if they don't have that young larva, then they are said to be hopelessly queenless. Really, a queen is a colony and a colony is a queen. If

Illustration 8: This is your first challenge to find the queen. She's bigger than the rest. Try to imagine them all moving. Photo credit: Marci Heiser

the queen is lost, then her genes are lost and even a replacement queen does not represent a continuation of her genes. The daughters she produced will eventually all die of old age and that is, literally, the end of that queen.

Now, have some fun trying to find the queen among the workers. First, you get a little help. Follow the gloved index finger.

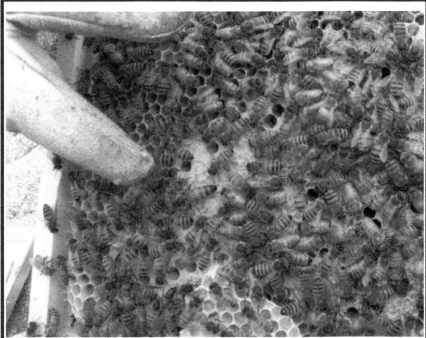

Illustration 9: Your second challenge with a little help, but not as much as the first. Remember, they are all moving! Photo credit: Marci Heiser

Illustration 10: This time you get put to the test! It's tricky. She's hiding in there. Photo caption: Marci Heiser

Illustration 11: Queen just to the right of the gloved finger with a large abdomen and shiny wings. Photo credit Marci Heiser

The answer for the previous picture is: she's that fat abdomen toward the lower left; partly buried under workers, to the right and lower than that glove tip on the left edge. Now a lot of help, just in case you're having a hard time. Try to imagine them all moving!

Workers are the brain, heart, soul and muscles of the colony. Workers also spend three days as an egg. They remain a larva for six days, being fed royal jelly about 1300 times each day on days four, five and six, but changing to a diet including **bee bread** and honey on days seven, eight and nine. They get **capped** at day nine. They spend 12 days as a pupa and emerge on the 21st day. They decide, direct and perform all the work. The queen is not the "boss" of the colony. In fact, she takes her orders from the workers. Workers decide how much brood the queen will lay, when and where. Workers decide when and if the colony will swarm. Workers will decide when and how much food the queen gets. Workers will carry away all the queen's waste. Workers will "run" the queen to make her lose weight if they are preparing to swarm. Together, the workers make up what is called the "**hive mind**" which makes all the critical decisions for the colony. Should they choose to swarm, the workers will organize and send out scouts to find a new home. Those scouts, which are all workers, will choose a variety of possible new homes and conduct a vote among themselves which will, nearly 100% of the time, result in them picking the best available new home. Their ability to unselfishly manage the colony absolutely makes human attempts at societal organization (capitalism, communism, socialism, fascism) embarrassing.

Illustration 12: Workers on comb with capped queen cells and capped worker cells. Photo credit: Kelsie Bell

Workers have to handle all these chores:

1. building the comb,
2. cleaning the used cells,
3. clearing out dead bodies,
4. foraging for water, propolis, nectar and pollen,
5. identifying a new home in the case of a swarm,
6. making and loading cells with bee bread,
7. making and loading cells with honey,
8. feeding the brood,
9. capping cells,
10. controlling temperature (92 - 95 °F),
11. defense,
12. feeding and caring for drones and the queen.

Worker duties are age and season dependent. Warm weather workers will live only 35 to 45 days, but cold weather workers will live several months. Upon emerging, workers are immediately **housekeeper**s, they must clean up their own cell to make it ready for a new egg. Summer workers will spend their first week performing housekeeping and feeding larva. They are frequently referred to as **nurse** bees. During their second week they will be building comb and receiving food from foragers. Their third week duties include clearing out corpses and performing **guard** duty, the most alert

Illustration 13: Broken queen cell with larva. Compare to drone and worker larvae surrounding. Photo credit: Don Studinski

bees in the colony. This is the time they are most likely to use a **stinger**. Should they choose to sting, their stinger stays in the victim (it is barbed) and the bee will die. After the third week, they will be foraging for water, pollen and nectar, literally working themselves to death. Winter workers are responsible for generating kinetic heat through vibration. That's pretty much all they do, so they can live much longer. Starting from the middle of the winter cluster, they climb up to get a bite of honey. Joining the cluster at the outer most point, they begin to shiver to create kinetic heat using up the carbohydrates in that honey. Slowly they work their way inward, always vibrating. Finally, they reach the innermost portion of the cluster where they are fully warm and can start the process again.

Illustration 14: Drone looking at you. Notice the big eyes compared to his sister just to the right. Another drone just entering the picture at the bottom. Again, big eyes. Photo credit: Don Studinski

Drones, male bees, are larger than workers, but smaller than the queen. Drones spend three days as an egg, eight days as a larva (capped on day 11) and 13 days as a pupa (emerge on day 24). They are a significantly different shape than their sisters, more stubby rather than long and thin. They have huge eyes and no stinger. They offer only one service: to fertilize a mating queen. Once they have given this gift, their organ is torn out of their body (similar to how a worker's stinger is torn from her abdomen once she has stung) and, consequently, the drone dies. Other than that, they are a drain on the colony. They have to be fed, kept warm or cool and take up space. That is, until fall, when those that have survived the growing season will be summarily driven out of the hive by their sisters. On their own, away from the colony, they will shortly die from exposure.

During warm weather, drones will make up about 5% of the healthy colony or much more for a queenless colony with a laying worker. If you observe an overpopulation of drones, it may be an indication of trouble and you should make it a point to verify the queen's health. Actually, you should make it a point to verify your queen health on a regular basis.

Honeybee Forage

Honeybees are either directly or indirectly involved with getting one third of the food on to your plate.[3] That is to say, without honeybees, your diet becomes much more repetitive and boring. In fact, honeybees are essential for pollination of 13 crops while significantly contributing to well over 100 foods we enjoy.[4] Not just exotic foods, but also many of our daily "take it for granted" foods like: onion, celery, broccoli, bell peppers, coffee, cucumber, squash, carrot, etc.

They haul home an average of 20-40 mg of nectar per trip[5] as their reward. They deposit the nectar in cells and fan the nectar with their wings to reduce the water content from as much as 80%[6] down to 18.6% or less. That's honey, our sweet, delicious incentive for beekeeping: nectar with moisture removed down to 18.6% or less. In fact, if you enter "honey" into a contest and it tests at 18.7% moisture, it's disqualified, that's not honey.

It takes approximately two million flower visits to create one pound of honey.[7] How can we know this? Here's the calculation:

---

3   http://www.ars.usda.gov/News/docs.htm?docid=15572 visited 9/23/2013

4   http://en.wikipedia.org/wiki/List_of_crop_plants_pollinated_by_bees visited 9/23/2013

5   https://insects.tamu.edu/continuing_ed/bee_biology/lectures/password/Internal_Anatomy_of_Honey_Bees_PN.pdf visited 9/23/2013

6   http://www.jackterry.net/honey/honey_factory.html visited 9/23/2013

7   http://www.ycbk.org/Honeybee%20Facts%20and%20Trivia.htm visited 12/21/2013

A pound is about 454 grams.

Assume 100 flowers per worker gathered load.[8]

Assume 32 mg nectar per load.

Assume, after dehydration, 32 mg of nectar makes 17 mg honey, the rest was water.

That's 100 flowers for 17 mg honey or 100,000 flowers for 17 g honey.

That means it takes about 2,670,588 flowers for 454 g or one pound honey

Illustration 15: Foraging on carrots. Carrots need honeybees to be able to make seed. Photo credit: Denise Conrad

How far will a honeybee fly to collect the nectar and pollen it seeks? Here's some math that may be interesting.

Typical[9] range is 2 miles radius = 3 km radius

Maximum[10] range is 7.4 miles radius = 12 km radius

Area = Pi(radius)^2

1 Square Mile = 640 Acres

Typical forage area:

3.1415(2 miles)^2 = 3.1415(4) = 12.6 square miles = 8042 acres

3.1415(3 km)^2 = 3.1415(9) = 28.274 square km

Maximum forage area:

3.1415(7.4 mi)^2 = 3.1415(54.8) = 172 square miles = 110,098 acres

3.1415(12 km)^2 = 452 square km

Given a diverse floral environment, honeybees will collect nearly double the pollen when compared to a monoculture environment[11]. From early summer to mid August, pollen is dominated by:

Fabaceae (example, soy, pea, alfalfa, peanut, clover)

Later in the year, pollen is dominated by:

Asteraceae (example, aster, daisy, sunflower, lettuce, artichoke).

8   http://www.honeycouncil.ca/chc_poundofhoney.php visited 12/21/2013

9   http://en.wikipedia.org/wiki/Forage_(honey_bee) visited 12/21/2013

10   http://www.ehow.com/info_8284470_range-honeybee.html visited 12/21/2013

11   Bee Informed Partnership, email 12/18/2012

February … Anticipating Spring

© steve kennedy 2014

Do you know the song "Anticipation"? I can't help it going through my head this time of year.

Which colonies will live? How many splits can we do? How many swarms will we capture? It's fun to dream of these things. This is certain at this point: some will live, some will be ready to split and there will be swarms to catch.

You really want to memorize the numbers below. The numbers promoted here are correct enough for you to perform your beekeeping with confidence. Exact numbers vary depending on numerous variables. Luckily, there is no need for you to spend time trying to remember the range of timing that might happen to raise a queen, worker or drone. Learning these generic numbers will be adequate for your needs. This factors into all your planning. Knowing life cycle numbers allows you to anticipate what's happening in the hive and, sometimes, that can save you from having to inspect. The good part about not inspecting includes:

1. Less work
2. Potentially less stings
3. Less bees killed by accident
4. Potentially prevent killing the queen by accident

Every bee goes through four stages: egg, larva, pupa, adult. Every honeybee egg is an egg for three days, then it hatches into a larva. Eggs hatch ... pupae emerge as adults.

Queen:	Worker:	Drone:
Larva: 6 days	Larva: 6 days	Larva: 8 days
Capped 9th day	Capped 9th day	Capped 11th day
Emerge 16th day	Emerge 21st day	Emerge 24th day

Illustration 16: A beautiful, calming, apiary in winter photo. Photo credit: Don Studinski

This time of year, beekeepers need to assess their colonies for health and population by observing them at the entrance. Any day offering you an opportunity to make an entrance observation while the temperature is above 50 (°F) gives information about how they are doing. Our objective is to know who is healthy and who may be failing. Failing is an emergency and may call for action. Generally, you want to leave that top on. This is "**no peeking**" season.

It's February 7, 2012, just a little more than a month from first inspection for our geographic area. The snow is deep which makes it nice to be inside. This is an excellent time to double check your inventory list to ensure you have everything you need and want for this year. If you don't have an inventory list, here's some suggestions:

- bee hive(s)
- extra supers for honey
- bee suit/veil/gloves
- **hive tool**
- smoker
- bees ordered

I also just got my catalog from Brushy Mountain Bee Farm. There is much in there to wish for, but I have to limit things to within my budget.

Illustration 17: Candy recipe cooked to "hard ball" by accident. Bees can eat it, but it takes much longer than usual. Photo credit: Marci Heiser

David found me a smoker container. It's a cute little trash can into which the smoker fits perfectly. This is going to be great for when I need to light the smoker at one location and try to hit other locations without it going out. I can pop the smoker into this metal can, put on the lid tightly and pop the can into my car and I'm off and running with a hot smoker and no fear of starting the car on fire.

This is a good time to identify any hive equipment that needs a fresh coat of paint. Haul that stuff into the house and paint it while you are wanting to be inside. Do not paint the inside of a hive. Leave the inside natural wood for the bees.

I like to name my bee colonies. This is not because I think of them as pets. It's because I do a lot of teaching and being able to refer to a colony by name when I'm communicating with students saves a lot of time over describing the position in the bee yard. In the remainder of this book you may see a colony referred to by name. There will be no confusion. The context will make it clear I am referring to a colony of honeybees.

The names I choose are related to the street where I captured the swarm or related to the vendor from which I purchased the bees.

Here's an example, I am not pleased with the eating progress on the candy in the Superior girl's hive. I think that candy may be too hard for them. When I made that candy, I mistakenly heated it to "hard ball" instead of "soft ball." I didn't want to waste the ingredients, so I decided to try it on them just to see what would happen. Although they are making progress eating through the brick, which you can see by the missing areas, their progress seems delayed by the extra hardness of the candy. Experiments like this are not uncommon when practicing permaculture style beekeeping. We learn as we go along. Next warm day, we will be putting a new block of candy on those girls.

Performing Postmortem

In the depths of winter, many a honeybee colony will perish. When this happens, the wise beekeeper inspects the frames of the dead colony seeking answers to the question, "What caused their death?" The most important thing to verify is that the cause was not a disease that could be transferred to other colonies. But, it's also informative to understand when death was potentially preventable and when it's just because it was their time. In order to identify the cause of death, inspect the frames and make observations using both your eyes and your nose. Do this in a timely manner soon after their demise. If you postpone, you may lose evidence to pests or deterioration. Here's some symptoms you may find.

Starvation:

The frames smell fresh. There may be frames with honey that the colony failed to move to. Many dead workers are present, heads down in the cells, butts sticking up in the air. A colony that fails to gather adequate stores or is unable to negotiate the cluster toward their stores due to temperature may die of starvation. Although it's never a happy end, there are times when it cannot be avoided.

Illustration 18: Students inspect frames left behind by dead colony. Photo credit: Don Studinski

Frozen:

Similar to starvation, this is where the cluster shrinks below the necessary volume for them to be able to maintain the temperature. The key difference here is that they may have stores immediately in their area and they have **brood** present, (new honeybees developing: eggs, larvae, and capped pupae). For example, a cluster the size of a softball may perish at temperatures in the teens degrees(F), whereas a cluster the size of a volleyball may not expire until temperature gets below zero. Somewhere there is a threshold where the cluster size is no longer able to produce enough heat. That's when they freeze to death. Of course, producing heat by shivering is dependent upon access to food. Further, the amount of heat required is affected by the size of the area to be heated. Some beekeepers are able to winter very tiny colonies even in very cold northern climates because the cavity size is appropriate for the small colony size and they are constantly in contact with their food.

In extreme climates, hives may require extra insulation wrapped around the outside.

Queen failure:

During late fall, given a substantial worker population, the hive mind (the collective decision of the workers) may choose for the queen to quit laying entirely. The remainder of the year, there should be some brood. When a queen fails to lay, for example in spring, the cluster begins to shrink due to normal attrition. When this happens, they are vulnerable to the temperature falling below what they can handle. This looks like frozen, but without brood.

Colony collapse (sometimes called **colony collapse disorder**, CCD):

This is not a disorder, nor is it a mystery. This is the normal result of a world covered with billions of pounds of poison every year. Bees are a signal of the ill health of our environment. Essentially, they go insane. Officially, CCD has these symptoms[12]:

1. Sudden catastrophic loss of the adult honeybee population.
2. Capped brood and plentiful stores of honey and bee bread are left behind which is not robbed by other honeybees. Scavengers show delayed interest in this food.
3. A few nurse bees and the queen remain.

These specific symptoms are harder to catch in winter because the temperature can also have pretty dramatic affects overnight.

Considering systemic poisons, like **neonicotinoid**s, one issue is in the bee bread. The girls have harvested pollen and mixed it with enzymes and honey to make bee bread. If that pollen is carrying a load of poison, then the poisoning happens upon consumption, which may occur in the winter months. Neonics are water soluble and probably do not find their way into the beeswax.

Considering other pesticides that are not water soluble, tests have shown as many as 39 unique pesticides and metabolites embedded in a single beeswax sample from a colony.[13] In this case, the concern is about adult bee exposure to the beeswax in general and larva exposure to the wax during development. Some hypothesize that the wax chemical cocktail may reach a threshold beyond bee tolerance resulting in CCD. However, I am not aware of any conclusive evidence to this effect.

Illustration 19: Five honeybees harvest corn pollen on one tassel. Can you find them? Photo credit: Bill Koeppen

As of 2012, 94% of all corn grown in the USA is seed coated with neonicotinoid poison.[14] Corn is wind pollinated, so it does not require honeybee assistance. However, wind pollinated crops, as part of their sex strategy, create a large amount of pollen. That's an easy and attractive source of protein for honeybees. As it turns out, honeybees love corn pollen.

---

12   http://en.wikipedia.org/wiki/Colony_collapse_disorder, visited 1/24/2014
13   http://www.panna.org/resources/panups/panup_20100326#2, visited 2/24/2014
14   http://www.panna.org/blog/bee-kills-corn-belt-whats-ge-got-do-it, visited 2/24/2014

As of 2012, 50% of all soy grown in the USA is seed coated with neonicotinoid poison.[15] Soy is not wind pollinated and does benefit from honeybee visits. Some pollinator visit is required for successful fertilization.

Varroa mites:

Colonies that die due to varroa overcoming the honeybee population will smell fine, but you will note abundant varroa mites on the bottom board. Look for tiny rust colored dots among the wax debris.

Nosema:

A spring disease, **nosema** is a tummy ache that can, but doesn't have to, kill. You will see honeybee stool on the front porch and, potentially, scattered over the hive bodies. This will be noticeably more than usual. The girls cannot control their bowels. More about nosema is covered in the chapter about diseases.

Swarm Preparations

During first inspection, coming up in March, we will be trying to get ahead of the bees' instinct to swarm. We want to prevent the swarms because swarms can end up choosing a new home where they are not welcome. We can prevent the swarm and leave the colony as a single colony, or, we can control the swarm instinct, using the split process, and end with multiple colonies.

Honeybees begin swarm preparation about five or six weeks before they actually swarm. If we find them "**honey bound**," that is, so much honey left that the bee population is getting crowded and they have inadequate space to expand the brood nest, then we can adjust the frames to free up space and reduce the likelihood that they will begin preparing to swarm. This is called **checkerboard**ing.[16] Your imagination can come up with what that looks like and it would be correct. We bring empty frames and replace every other honey frame with an empty frame. Three empty of ten frames should be adequate. We are talking about honey frames here,

Illustration 20: Many queen cells shaped like peanuts. Capped drone cells surrounding, bullet shaped. Photo credit: Marci Heiser

not brood. Honey left over after winter will be in the top box or two, depending on how far up the brood nest has moved. Do not checkerboard your brood.

15 http://e360.yale.edu/feature/declining_bee_populations_pose_a_threat_to_global_agriculture/2645/, visited 2/24/2014
16 http://en.wikipedia.org/wiki/Checkerboarding_%28beekeeping%29, visited 12/24/2013

Let's suppose you are using 10-frame equipment with two deeps. Let E represent empty, H represent honey and B represent brood. Your two stacked deeps would look something like this after checkerboarding.

HHEHEHEHHH
HHBBBBBBHH

Done early enough, this will prevent swarming without fail. Additional space can be provided by stacking honey supers on top when the flow arrives. We learn this in February so we are ready to use this knowledge in March.

Here's what honeybees do to prepare to swarm:

They begin by expanding the brood nest and laying a lot of eggs which will become workers. They want a large population of young workers. Most of these bees will fly with the old queen on swarm day. This is because the young workers are the ones that can make wax. The new colony will, most likely, need to make comb from scratch when they arrive at their new home. As these new workers emerge, the colony begins to back-fill the brood nest with honey and bee bread. They shrink the brood nest. This is because, once half the bees fly away with the old queen, those that remain have a reduced nurse bee population for warming the brood. Further, because their population is reduced, they also have a reduced forager work force which will slow down the incoming nectar and pollen.

While the brood nest is being **back-filled** with plenty of honey and bee bread to get the old colony through their transition to a new queen, they pick a young larva as their new queen and continue feeding her royal jelly. At this point, they are committed to swarming. This is approximately nine days before they will fly. Once they have crossed this line, it's very difficult to get them to stop. They feed a normal worker larva royal jelly about 1300 times each day from day 4 to day 6. A larva intended as a queen will

Illustration 21: Nearly perfect location. By shadows, you can tell south is to the left. Wind break is to the west and north. Creek is at left edge of photo. Photo credit: Don Studinski

continue to be fed royal jelly an average of 1300 times during days 7, 8 and 9.[17] This is what causes the queen to get a fully developed reproductive system. They will be capping her highness' cell on the 9th day. Frequently, there will be multiple queen cells. The bees take no chances on failure. Capping of the queen cell is the signal for the swarm to prepare to leave.

17 Personal correspondence, 12/20/2013

The workers will "run" the old queen to make her lose weight in preparation for flight. The old queen must be gone before the new queen emerges (on day 16) or the two queens will fight to the death which would be counterproductive. On swarm day, preferably warm with no wind, the old queen and approximately half of the workers, heavily weighted with young workers, will fly. They will land within about 100 yards from the mother hive and they will hang out there while the scouts pick a new home. The lucky beekeeper to find them in that state will grab them. It's a free colony of bees.

Meanwhile, back at the mother hive, the new queen emerges about day 16. She spends some time settling in. She must sexually mature and then perform her mating flights. This whole process takes about 2 weeks. Once that is complete and she has safely returned, she begins laying and will not fly again until (and unless) the colony decides to swarm. There is one exception, of course. We call that abscond. Abscond is quite rare.

Hive Placement

If you will have a living colony this year, you should be preparing your hive now. Select a **location**. Consider wind. Because we live along the Colorado front range, you want a wind break to the north and to the west. That is where our prevailing winds can cause trouble. Consider access to water. Will you water them every day like a dog or will you allow them to access a creek or a pond or a lake? Consider sun. I prefer full sun, however, that preference is evolving as our climate warms. A deciduous tree to the south is becoming increasingly attractive as we experience more 100-degree days. Prepare a hive stand which will keep the hive six inches off the ground. This eliminates rot in our climate and reduces pests invading the hive. Level your hive stand east to west. Tilt your hive stand slightly to the south. Front door placement is to the south and you want rain water to run out of the hive.

Illustration 22: Found queen, moved her to nuc. She will be moved more than two miles. Remaining queenless hive must make new queens. Note date on photo. Photo credit: Don Studinski

Planning Splits

Believe it or not, February is not too early to start your planning for splits. The keys for planning splits are drones and weather.

In Colorado, I've seen a first capped drone on 3/12/2011 (capped on the 11th day). That means, at the least, drones will be flying and available for a virgin queen 13 days later on 3/25 (emerge on the 24th day). So, theoretically, I could start a split on 3/9 giving 16 days to create a queen that would emerge on 3/25 along with the drones.

However, that is too soon, in my area, due to weather. And, I want to ensure the drone population is plentiful. We are still having consistently cold nights in March along with substantial snow (our snowiest month). So, my plans are for a 3/25 split date with 4/1 as backup if weather will not allow 3/25. Still, the truth is that the bees know better than I do. Watch for abundant capped drone cells. That's the bees telling us that mating time is near.

Given a 3/25 split, I'm looking for the new queen cell on 4/1, week 1. This inspection is particularly delicate. I must not damage the queen cells. They will be sticking out between the frames. If I have good cells, then I complete the splits (every new colony must have honey, bee bread, nurse bees, brood, queen cells) and start a no peeking wait time. I'm expecting the new queen to emerge about 4/8 (that's Easter in 2012), week 2. She needs 2 weeks to get acquainted with her new home, sexually mature and take her nuptial flights, weeks 3 and 4. She will mate with between 10 and 20 drones. Each of them will give their life for the privilege. She will be home and ready to lay about 4/22 (that's Earth Day).

My eyes are not very good, so I cannot see eggs. Therefore, I will wait another week before peeking, 4/29, week 5. I want to be able to see freshly capped brood. Eggs laid 4/22 will be ready to emerge approximately 5/13, 21 days later (that's Mother's Day). The bee population will grow rapidly from that point forward. That gives them about a month before the summer nectar dearth when blooms containing nectar become harder to find.

If I don't find fresh larvae and capped brood on 4/29, then this split has failed. In this case, our options are to merge them with a **queen right** hive, preferably their sisters (where they originally came from) which would readily accept them, or order a queen from a vendor. Since we are doing spring splits, queens are generally available for purchase.

Illustration 23: Drone frame. Photo credit: Don Studinski

Drone Frame

The green frame in the photo on the previous page is a **drone frame**. Here's a close up of that frame. I experimented with drone frames in 2011 in an effort to reduce varroa mites, but decided using them is not for me. The beekeeper battle with varroa mites is covered later in this book. Drone frames are intended to help with that battle.

The strategy with drone frames is that if you get the queen to lay a large drone area all on one frame, then this will attract the varroa mites within the hive to that particular frame. Varroa are known to prefer drone larvae and they are attracted to that area by a unique smell. The varroa will have laid their eggs on those drone larvae and the adult mites may then leave and end their natural life-cycle.

Once those drone larvae have been capped, all those newly developing varroa mites are inside those capped cells. A beekeeper that is really on top of things can take that frame of capped drone out sometime between day 11 when they are capped and day 24 when they will emerge. Removing that frame within that time window means you have removed the majority of your "next generation" of varroa mites. Put the frame into a freezer over night and you have killed all the developing varroa mites. You have also killed all the honeybee drones on that frame.

The next move is to put the frame back into the hive to allow the bees to clean out the dead drone bodies along with all the dead varroa mites. This frees up the cells to begin the cycle again.

There are several reasons I didn't care for this approach. First, I lost all those drones and because I breed my own queens, I actually want a large and healthy drone population around my hives. Second, all those dead drones represent a large workload for my undertaker bees which must remove the dead. Third, if you miss the timing and allow those drones to emerge, then you have just increased your varroa mite population much more than a normal portion of drone comb would. And lastly, it's quite labor intensive and my family doesn't appreciate honeybee brood combs in the freezer.

As you will read later, there are better ways to live with varroa mites rather than inventing ways to kill them.

Exterior Hive Seal

Marsha writes:

I don't like the paint look. Is there a natural oil or sealant that could be recommended instead?

Answer:
You don't have to be a beekeeper long before you find out that for any given question, if you ask 4 beekeepers, you will get 5 different answers. The question was "what shall I use to coat my hives to preserve them if I don't want to use paint?" I've gotten a lot of possible answers from my beekeeping communities.

1. Beeswax them! A hairdryer helps to play out the wax.
2. Beeswax will cause you major grief on the hot summer days.
3. Use barn and fence paint. White lids and red hive bodies.
4. I have 30 year old unfinished exterior grade plywood that is still structurally sound. In this climate (Colorado), as long as you keep it out of the dirt, wood does not rot very fast.
5. Milkpaints of old.
6. Boiled linseed oil is a popular solution, linseed oil requires periodic re-doing.
7. You can use any kind of oil you want on the outside. Linseed oil is often used but I don't think it's very durable.
8. Approx. 17 coats of Homer Formby's hand rubbed finish.
9. 1 litre linseed oil (raw or boiled, it matters not as you are about to boil it anyway) add 50 ml melted beeswax (use 1:20 ratio with whatever units suit you). Heat in a double boiler (bain marie; or one saucepan inside another – the larger one containing a couple of inches of water). Get it as hot as boiling water will allow and stir for 10 minutes. Allow to cool and while still on the hottish side of warm, paint it on the outside of your hive, paying special attention to end grain, nail heads (underneath) and joints.
10. I mix some old wax in with linseed oil and apply that. The oil helps carry the wax into the pores, and the wax gives a lot more protection than the linseed oil. One of the biggest

advantages is that in re-coating you don't need to sand like you would with any of the film finishes like paint or varnish. Oil is a penetrating finish.

11. All oil base products are detrimental to insects.

12. I use marine varnish, like what is used for boats. I like the oil based products more than water based products because I think they get soaked up by the wood instead of being just on the surface.

13. Some bee equipment manufacturers are using cypress for bee hives with no finish. Check out Brushy Mountain Farm.

14. Santa Fe beekeeper who has used it and likes it: www.ecowoodtreatment.com

15. Info about dipping in paraffin / wax: http://www.bushfarms.com/beeslazy.htm#stoppainting

16. The only substance that's colored I know of that cures (thirty days) into an inert substance would be 100% silicone which also breathes.

Okay, see what I mean? This is a real list gathered from real beekeepers. It specifically recommends oil and specifically says don't use oil; specifically recommends paint and says don't use paint; etc. From a permaculture perspective, we should be seeking solutions that require less work and stay more in tune with nature's way. My recommendation is do nothing. Let the wood age. It's beautiful.

Interior Hive Seal

Do not paint the inside of a hive, or any part of the inside, with anything. Leave it natural wood.

Illustration 24: See the propolis envelope. It gives the wood a darker color inside the hive's false back. The arrows show the line. The arrows sit on the naturally aged wood and point to the propolis envelope coated wood. Photo credit: Don Studinski

Propolis is tree resin. Honey bees collect it intentionally. They are medicinally treating their home. Propolis has been scientifically shown to be antibiotic, antiviral, antimicrobial and anti-fungal. Furthermore, scientists have found that the resin collected changes based on what is happening with the health of the colony. The honeybees are acting as their own pharmacists.

You have probably noticed that honey comb is much softer and more delicate than is brood comb. You have also noticed that brood comb is darker, sometimes very dark. The reason for the darker color and thicker walls of brood comb is, in part, because the bees are coating the cells with propolis before each use. This makes each cell a nice secure environment for the egg, larva and pupa to develop. The other reason for the darker color and thicker walls is the cocoon left behind after each metamorphosis.

Beyond coating the comb cells, honeybees surround themselves with a **propolis envelope** throughout the hive. You can see this in the photo.

Here, we have a top-bar hive from which a false back has been removed. The wood has been weathering for three years. This reveals a very visible color difference between where the bees have been covering the wood with a propolis envelope (inside) and where the bees could not reach (outside the false back). Propolis provides a health shield for the bees. Some beekeepers have noticed that the bees provide a better propolis envelope on rough surfaces than on smooth. Therefore, some are now roughing up the interior of their pine Langstroth boxes intentionally to encourage the bees to make a nice solid propolis envelope.

Early Pollen

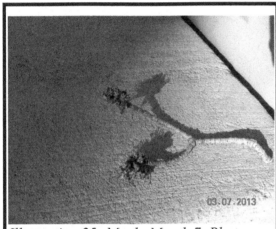

Illustration 25: Maple March 7. Photo credit: Don Studinski

Even in March, honeybees are already finding fresh food sources. They notice and take advantage of assets we hardly even know are there. I start noticing workers

Illustration 26: American Elm March 7. Photo credit: Bobbi Storrs

hauling in pollen very early in the year. Curious about their source, I started heightening my awareness. I found these tiny blossoms on a Maple tree. You can hardly see them. I've tried to capture them using shadows. Bobbi found similar blossoms on an American Elm.

Know Your Cells

April of 2011, I got some good photos of frames which clearly show each cell type beekeepers need to learn to identify. This chapter includes large photos. This has an upside and a downside. The upside is that you can easily see what you need to learn. The downside is that the text is not always on the same page as the photo being described. You will have to do some flipping back and forth to get the best experience.

Although many readers may already have this knowledge, it is still fun and instructional for the casual reader and those that may just be thinking of becoming beekeepers. In every photo, you will see some **empty cells**. Pale yellow cells are delicate to the touch and easily broken. When you see that pale yellow color, it's either newly built comb, or comb intended to hold honey. Bee bread and **brood** will be stored in the darker colored cells where the interior of the cell has been coated with propolis.

Illustration 27: Queen cell, worker brood, honey to the corners. Photo credit: Marci Heiser

Starting with the easiest cell, **capped honey**, you can see examples of this in the upper corner of the frame shown on the previous page. Note that the upper corner of the frame is being held by my left hand in this photo. The bottom of the frame is shown vertically to the right in the photo. The honey is under those pretty white wax caps. Note also that this area has some empty cells where the bees have been consuming the honey. Those cells are likely to be refilled with honey as spring progresses. Now look at the opposite corner, the upper left corner of the frame, toward the bottom of the photo. More capped honey. This is a typical pattern you will see the bees using: honey in the upper corners of a brood frame.

In that same photo, under my right hand, is the brood area of the frame. This is a nice football shaped pattern indicating a good and healthy queen. All those brownish capped cells are female brood, **capped workers**, which will emerge for the 2011 season. All the bees you see in this photo are females, workers, probably "nurse" bees.

I'm pointing to a particularly unique cell toward the bottom right of that frame. Take note of that cell. It is a great example of a queen cell. There are some "usual" aspects to this cell and some "unusual" aspects to this cell. The usual aspect is related to the shape of the cell, a peanut shape. This is a typical shape and size for a supersedure queen cell. The term supersedure cell is reserved for when the colony has decided to replace their queen. This cell type is frequently found in the upper part of the frame. This one is on the lowest portion of the frame. That's what makes this unusual.

Now you are probably thinking, "Hold on there, you just said the brood indicated a healthy queen, but then you said the bees have decided to replace her!" That's right; good catch. What I was holding out on you was that I had deliberately removed the healthy queen a week earlier. I essentially forced a strong three deep colony to be queenless. I took their queen, a couple frames of brood and a couple frames of honey and made a split. I loaded all this into a "nuc" and took them to my apiary several miles south. This was essentially an artificial swarm. Being left queenless, the three deep colony was forced to make a new queen. We call this an "emergency" queen cell. They had to pick a larva in its fourth, fifth[18] (according to some) or sixth day of development[19],[20],[21] (according to others). Because this was an emergency, it appears that the "right" larva was toward the bottom of the frame. Sometimes, bees will decide to supersede their queen on their own. When this happens you may find the queen cell on the upper portions of the frame.

Lastly, notice the **queen cup** just above the queen cell. That kind of cup is where a new queen could be raised in anticipation of swarming.

For the photo on the facing page, notice that this is a full brood frame. No honey in the corners on this one, though there is some light colored cells in the corner that could be used for honey. If you look closely, you will notice empty cells in the upper left and along that left side which are slightly bigger than cells containing brood on most of the frame. There are also some slightly bigger empty cells in the area I'm pointing. Those are likely candidates for an unfertilized egg. That's how a drone gets made, the queen would lay an unfertilized egg into those slightly larger cells. I'm actually pointing to two capped drone cells. They are larger, a lighter color and poking out slightly higher than the rest of the brood cells (sometimes called a bullet shape). There's a bee crawling over the higher one. The rest of this frame is capped brood, female bees, workers.

18   http://scientificbeekeeping.com/fat-bees-part-1/, visited 8/1/2014
19   http://en.wikipedia.org/wiki/Royal_jelly, visited 8/1/2014
20   http://www.uky.edu/Ag/Entomology/ythfacts/4h/beekeep/beebio&s.htm, visited 8/1/2014
21   http://westmtnapiary.com/bee_castes.html, visited 8/1/2014

Illustration 28: Capped drone cells among capped worker cells. Photo credit: Marci Heiser

The frame in is covered with a large, healthy number of workers. Looking very closely, near the dead center of the photo is one bee that is noticeably larger than the other bees. He's looking down and to the right with a dark butt sticking up and to the left. That's a drone. The bees don't make drones until they are "ready" for swarm season. That's when drones are useful. Drones don't have stingers. Their job is to fertilize a virgin queen. In fact, that's their only job. They fly to a **drone congregation area** (DCA) approximately 50 feet in the air. There, they hover in wait of a virgin queen. When the virgin arrives, the drones compete for her affection. If they win, and they get to mate with her. Their organ is ripped from their abdomen and they die after they mate. If they don't win, then they can return to a hive at dusk. Nearly any hive will do. Drones are readily accepted for entry, even when it's not their home. This is a mystery because, while they are there, "all they do is smoke cigars, watch TV and drink beer." The workers actually feed them and control their temperature.

It seems like it would prevent the spread of parasites and disease if the drones could not so easily drift from hive to hive. But, for some reason, nature has chosen to set it up this way. It could be to help spread the genes farther as drones drift away from their original home.

Illustration 29: Brood frame back-filled with honey. Not surprising 4/21. Photo credit: Don Studinski

This is a brood frame where the bees are back-filling with honey. You can tell this is a brood frame from the dark colored empty cells visible toward the bottom. The capped honey toward the bottom is called **wet honey** because the honey is actually touching the wax cappings. There is also new nectar glistening in cells toward the bottom. That's where the bees are actively working during this photo. The capped honey toward the top, which is white, is just called capped honey. The big chunk of comb which looks out of place at the top is called "**bridge comb**" because the bees built it as a bridge between this frame and the next. It was attached to both and had to be broken to get the frame loose. This was found in the brood area of a hive 04/21/2012. You can learn about why the bees back-fill the brood area with honey in Anticipating Spring, Swarm Preparations.

Illustration 30: Old brood comb ready to melt after the brood emerge. Photo Credit: Marci Heiser

This is a brood frame. It's not exactly full of brood. In this case, that is not a problem. This frame has a lot of empty cells where the girls have already emerged. This entire frame was used exclusively for brood. What I want you to notice about this frame is that it's time to replace it. See that dark colored wax? This frame is several years old. It's likely got pesticide residue in it. After all the brood have emerged, I'll pull this frame out and melt the wax. I don't want my frames to hang around too many years. I want my girls to stay healthy and new wax is a comforting warm-fuzzy in that respect.

March ... The Most Exciting Time of Year

© steve kennedy 2014

Excitement is in the air. The spring equinox is just around the corner. Queens are laying in earnest as the spring nectar flow approaches. Soon, nectar and pollen will be abundant, more abundant than at any other time of year. Will this be the year for that bumper crop of honey? Will this be the year the bees are so healthy you are able to triple your livestock? Soon, the answers will be rolling in. Who can resist being excited?

Swarm season is almost upon us. Final preparations need to be made to collect and place the new arrivals. This means hive equipment is in place and ready to receive a new colony. Nucs are prepared and placed where needed. The swarm team is all arranged and they are ready with proper clothing and swarm containment equipment in their vehicle.

Packages and nucs have been on order for a while. It's time to finish up preparing their destination hive.

Dead-Out Inspection

The Longmont girls are dead. David notified me yesterday, 2/29/2012, that the hive had no activity on a warm day so he peeked. This gives us an opportunity to learn about combs, hive layout, dead hive postmortem, what to do with the equipment and why the girls sometimes don't make it even though they ALMOST made it to spring.

We took a careful look at every frame of the dead colony, 3/5/2012. We reconstructed the brood nest and surrounding honey so we could examine what the bees were doing and why. We learned about what killed them and a variety of potential killers that were not present in the evidence we could see and smell. Some lucky students brought home a Tupperware container of honey. We learned about worker cell size and drone cell size. We also examined a queen cup. We learned about cappings and why some are darker than others (hint: white capped honey has a small amount of air just below the cap). I am confident that what killed this colony was a prolonged cold spell for which they were not able to negotiate their hive **topology**.

Honeybees will cluster whenever the temperature drops below 57°F, though I frequently see them flying even slightly below 50°F. By gathering together into a tight formation and shivering, they create kinetic heat to keep themselves alive. We call that a cluster. Workers take turns starting from the middle of the cluster where they are fully up to temperature so they can move around easily. From there, they venture out to the food to get a bite, and with a full tummy, they rejoin the cluster at its outer-most mantle. There, they begin to shiver to create heat. From that point they slowly work their way back into the center of the cluster and start the cycle again.

For this dead colony, the brood, which they <u>must</u> keep warm, was too far from the food. They will not leave their brood. Therefore, with maybe 15 lbs of honey easily available, they starved. This was not disease, not poison, not the available food. It was the prolonged cold which prevented them from moving to the food.

What could have been done to get their food closer to them so they would have been able to get to it? If it's too cold, is it better to open the hive to get their food to them or just hope for the best? I am not aware of a way to move the honey toward the brood without anticipating the problem and making that move on a warm day when you can open the hive. You can move it over next to the brood nest, but you must not split the brood nest. It's a hard call in winter when I'm sticking to "no peeking." I generally stick to the "hope for the best" strategy. I trust my bees to know what's best for them. In this case, I lost.

Illustration 31: Reconstructed brood nest. Photo credit: Ruth Rinehart

The reconstructed brood nest, in this case, is three frames in the middle, surrounded by honey. Brood nest in the middle with honey to the sides and above is the standard configuration for honeybees. Notice the bee flying by the blue tub. This is about two miles from where the dead-out occurred and we are not aware of any hives in the area, however, honeybees can find you very quickly whenever they smell honey.

Notice the larger drone cells at the bottom of the empty brood frame. The rest of the frame is worker brood size. The cells with brightly colored spots in the bottom contain bee bread in an unusual, widely scattered pattern. It makes me wonder what they were thinking.

Bees frequently create queen cups even at times when they have no plans to replace the queen. This example queen cup was right in the middle of the frame just above some missing comb. Queen cups intended for swarming are more commonly found along the edge of the frame, especially toward the bottom. This cup in the middle takes a position similar to a bottom edge because of the comb missing just below it. The bees regularly prepare queen cups so they are

Illustration 32: Empty brood frame. Photo credit: Ruth Rinehart

Illustration 33: Queen cup. Photo credit: Ruth Rinehart

readily available whenever needed. The presence of a queen cup doesn't necessarily imply that the colony is preparing to swarm. It just gives them the option. Above the queen cup is capped wet honey. We call it wet honey because the honey is touching the wax caps. When the honey has a small space between the liquid and the wax cap it leaves the wax a beautiful white. A whole frame of evenly built white capped honey can be a prize winner at the fair, but wet honey is doesn't fetch a blue ribbon.

When we take a look at the cluster as they died, we notice the butts in the air of girls licking the bottom of cells for the last of the food, despite the fact that honey is close by. You have to look closely to see bees in the cells with the tip of their butt just poking out. They could not leave the

Illustration 34: Dead cluster. Note capped honey immediately next to bodies. Also note butts in the air. Photo credit: Ruth Rinehart

cluster where the brood was being raised. They may also be cannibalizing some of the larvae. These bees likely froze to death because we see they had food within reach and brood in progress.

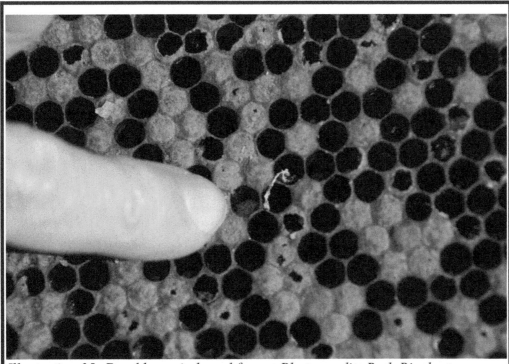

Illustration 35: Dead larva in brood frame. Photo credit: Ruth Rinehart

In the dead larva photo, you can see a dead larva in the bottom of the cell. These golden brown capped cells were raising brood at the time of death. Some of the brood were just emerging as they died; those are the ones with holes in the caps. The tiny holes in the caps may also be evidence of varroa mites emerging after the larva under the cap has died.

It was a sad thing,

but it happens. Beekeepers must deal with dead colonies. On a happy note, there was no disease. On an even happier note, the students got treated to some fresh honey. It's liquid gold.

Illustration 36: Liquid Gold. Photo credit: Ruth Rinehart

First Inspection

March is first inspection time in the Denver Metro area. No peeking season comes to a close and we get to find out who is alive and who didn't make it through the winter. Planning for first inspection has been underway for some time; now we execute our plan.

The determining factor for first inspection is temperature. You must have temperature above 50°F and we prefer no wind. We also target four to six weeks ahead of first swarm. We are trying to get ahead of the swarm instinct.

We combine first inspection with some extra tasks that are meaningful for this time of year and reduce the number of times we disturb the bees which is good for them and easier on us.

1. Perform checkerboarding for swarm prevention on any honey bound hive.
2. Assess for split potential, especially strong colonies.
3. Perform spring cleaning of all hive equipment

If we find a colony honey bound, that is, they have no room to expand their brood nest because they have too much stored honey left over, then we want to use checkerboarding to give them the added room they need. Checkerboarding[22] is when you remove alternating frames of solid honey and replace them with frames of drawn comb or **foundation**. Replace three of 10 frames in the box over the brood area. You may also consider adding an additional super over the box where you do the checkerboarding. Using this technique, you will prevent a swarm.

Shortly after first inspection, we will execute our split process. Colonies that are strong, being quite likely to swarm, can be split to multiply our honeybee inventory. This is sometimes referred to as **increase** in beekeeper vocabulary.

Illustration 37: Abundant Drone Cells. Photo credit: Don Studinski

---

22   http://www.knology.net/~k4vb/ABJ%20Copies/BC%20Mar%201997.pdf, visited 2/26/2014

The keys to successful splits are drones and weather. When capped drone cells appear, the bees are saying, "We are ready for virgin queens." We could anticipate the drones and try to time our splits to coincide with new drones emerging, but since we want <u>abundant</u> drones, rushing is foolish. By waiting for abundant drones, we reduce our vulnerability to changes in timing that the bees know about, but we would mess up.

Illustration 38: Light smoker. Photo credit: Marci Heiser

I do not use drone sheets or any drone stimulating equipment not produced by the bees themselves. I want the girls to tell me when the time is right for splitting by naturally producing males. I try to avoid introducing my own human bias which may influence this process in a negative way. I need to be able to count on feral colonies living in tree hollows also making males at that time so my new virgin will have plenty of appropriate mates from which to choose.

Once I've spotted abundant drone cells, the only question left is, "How's the weather?" If we are having a particularly cold spell, then I delay splits. Otherwise, I'm ready to go. This typically comes toward the end of March or the first of April.

Illustration 39: Taking Notes. Photo credit: Don Studinski

I keep bees in a variety of equipment types. Splitting is <u>much</u> easier when the source hive is the same type as the destination hive. This is because you want to be able to lift brood comb from the source and place it into the destination without disrupting the comb, or the nurse bees. For example, I'll split Langstroth to Langstroth or top-bar to top-bar (same frame dimensions) whenever possible. I can split top-bar to Langstroth, but this requires special transition equipment for the TBH frames to fit inside the Langstroth box. As you are planning for splits, you will want to place the extra equipment near where it will be needed and used. It's not too early to get your equipment in place now.

Illustration 40: Note Taker. Photo credit: Don Studinski

We'll cover the split process in more detail in April.

I'm motivated to disturb my bees as little as possible. This is partly because of the fact that they are a **superorganism**. Essentially, inspecting a hive is like performing surgery on the superorganism. It's very disruptive. You cannot open a strong colony without killing bees. Think of how difficult it is for a body to recover from surgery. Healing takes significant time and energy. I prefer to save that energy for honey collection and on-going hive health maintenance rather than forcing recovery from an unnecessary surgical procedure.

The main objective of first inspection is to see and make notes on the

Illustration 41: Screened bottom board can make a handy smoke entry point. Photo Credit: Marci Heiser

Illustration 42: Capped honey, bee bread sphere, larvae, capped larvae. Photo credit Don Studinski

health and well being of every frame of every hive. We break the propolis seal and go inside, all the way. We use smoke and there are bees in the air. We will not rearrange their brood nest. We leave that as we found it. However, if we find them honey bound, then we will be harvesting some honey and checkerboarding frames over the brood nest.

Think of the brood nest as a volleyball shape in the middle of all those rectangular frames and boxes. That's where the new bees are growing. Immediately surrounding that is a sphere of bee bread: pollen mixed with enzymes and honey (this is their protein). It's kept here so it's handy for the nurse bees to eat so they can produce royal jelly and feed both the royal jelly and the bee bread to the larvae. Outside that sphere is honey up to the corners and edges, as far as the bees can build within this rectangular hive. If you keep this shape in mind, then when you see an arch of bread-filled cells over the brood area on a frame, it begins to make sense. You can see the bee bread in the photo above. The capped brood is toward the center of the image, left of that is a portion of the bee bread sphere. Those cells are loaded with brightly colored pollens. Left of that is the capped honey to the edge. You can see uncapped larvae in this image.

Illustration 43: Mama Birch shows how to get ready for the spring flow. Photo credit: Marci Heiser

When you pop open a hive mid March and find 8 frames of bees, it is a very joyful moment. These Birch girls immediately go on our "to do" list to be

Illustration 44: Mouse damage. Mouse ate comb, brood, honey and bees. Urine and feces everywhere. Photo credit: Don Studinski

split very soon. In fact, this colony ended up getting split 3 ways. Mama Birch moved to Golden. Both daughter queens stayed here, in Broomfield. They all performed very well and produced some harvest in 2012. The Birch line was one of my best performing for several years. I was always glad to get daughter queens from the Birch line.

Sometimes, what you find at first inspection is not happy. In this top-bar hive (TBH) we found that a mouse had eaten comb, honey, all the brood and generally made a mess of the hive with urine and feces. The damage had just happened recently. I had been watching this colony closely all winter. They were generally touch and go, but had just blossomed in a huge way with a lot of activity and many bees. After having successfully survived winter, these girls were dead within two weeks of first inspection.

Remember, rodents carry disease, especially where they leave droppings and urine, which is everywhere you see evidence of them. When you find yourself having to clean up such a mess, be sure to wash. Wash gloves twice with soap. Scrub each finger of each glove as well as palm and back. Wash your suit in the bath tub. Don't forget to wash your hive tool and any other tools you used to clean up the mouse mess.

Illustration 45: European style hive. Photo credit: Don Studinski

In 2012, we also got to see Bill's BeeHaus. As it turned out, the few bees in here were **robbers**. These girls had died of cold. This was another case of plenty of food available, but not close enough to the cluster to save them.

First inspection will be the smallest population of bees you encounter all year. Therefore, take this opportunity to scrape **burr comb**, clean the bottom board, swap out any deeps that need a coat of paint (remember we just put new paint on deeps that were resting over the winter). The entire hive will be disassembled, so this is the ideal time to clean up. When the weather is right, open your hive (use smoke) and inspect throughout one frame at a time. You should see plenty of brood (female) and you are hoping for abundant brood (male) as an indication of when it's time to start splits. Scrape burr comb as you go along if you wish. If you run into broken frames or boxes, this is an excellent time to rotate out the broken and rotate in the replacement. Disassemble the hive all the way to the bottom board and scrape that bottom clean. If the colony is suffering from nosema, swap out the soiled bottom board for a clean one. Reassemble the hive with the unbroken and clean equipment. Take the equipment needing repair and cleaning with you as you leave.

How might you recycle that broken equipment to get more use of it? For example, a 10-frame deep with a broken corner might be cut into a few **candy super**s. The bad corner is eliminated, and you still get good use out of the remainder. Equipment that is truly at end of life may be best burned, just to ensure that there is no disease or parasites left to spread.

An injection of common sense: if, for some reason, the weather is not ideal, then you need to hurry through first inspection. Skip the clean up stuff. Save your bees.

Illustration 46: First Quay Street Swarm. Notice the Y shaped branch just behind the cluster. Photo credit: David Braden

More About Swarm Cycle

Bees start to gradually expand their brood nest after the winter solstice. The daylight gets longer with each passing day. This practice will continue based on available food stores, workers (which are necessary to warm the brood) and space. Food stores include what they have available in the hive which they had stored for winter and pollen and nectar flow that begin late winter and early spring. Space means empty worker brood cells. Having gradually expanded the brood nest since solstice, the bees are assessing their ability to swarm by about the first of March. If they are strong enough (they've successfully raised enough workers) and have adequate stores, they will swarm. It's their instinct to reproduce, a beautiful, happy thing.

Case 1: They decide <u>not</u> to swarm. Brood rearing continues without interruption as they continue to expand population in anticipation of the spring nectar flow. This may be a good colony for honey production.

Case 2: They decide to swarm. Brood rearing declines! They start freeing up brood space for nectar and bee bread. The worker population is already adequate for swarming. All that's left is to allow a bunch of young bees to emerge so the swarm will have plenty of workers who are ready to do wax production. As they emerge, they save the empty cells for food, not brood. Because after the swarm, the remaining bee population will not be as able to generate incoming food. They will be coping with a reduced forager population for a while.

Illustration 47: Second Quay Street swarm. Next day, but the same spot as the first Quay swarm on the same branch. Photo credit: Don Studinski

The swarm cycle begins a month to six weeks in advance of a swarm. That decision has been made when the brood nest begins to reduce in size. It's almost like a big breath "in" from solstice to maximum brood nest size, followed by a big breath "out" from maximum brood nest size to swarm. The bees "commit" to swarm when they start rearing new queens in swarm cells. But this happens only about one week prior to swarming. They actually began their preparations at least three weeks prior to beginning a queen cell. Once that queen cell is capped (on the 9th day after the egg was laid), the old queen will be run off and she will take approximately half the bees with her, mostly young workers. That's a swarm. They have a few days of grace for weather issues, but she must fly soon after the new royalty is capped. Only seven days remain before the new virgin queen emerges. If the old queen is still there when her daughter queen emerges, a fight is coming.

To effectively prevent a swarm, you must be ahead of those early preparations. That's why we use checkerboarding at the time of first inspection. Any colony that has too much honey and not enough room will be checkerboarded: three frames of honey come out of the upper-most deep, replaced by three frames of empty comb, alternating / honey / empty / honey / empty.

This eliminates the swarm trigger. They have plenty of room for expansion. They feel no crowding. They choose not to swarm. This is great for us because a swarm is an uncontrolled event. We want to control that event by performing an artificial swarm. We call that a split. We still want them to multiply, but in a controlled manner.

Illustration 48: Pecos Swarm. Photo credit: Don Studinski

We can prevent swarms in **Warré** hives exactly the same way. Top-bar hives are a bit different. In this case, consider a substantial harvest. Give the bees as much room as possible where they will be encouraged to build new comb and potentially migrate their brood nest to the new comb.

Remember This in March

I can't wait to see my bees. As abundant food draws near, the bees relax and become very easy to work. We may not be quite there yet, but we are very close. When you see photos of people bare skinned, covered with bees, wearing them like a coat, it's a spring thing. They may not mention it, but it is a spring thing. From spring equinox

03.10.2012

Illustration 49: Birch, first inspection, 2012. Photo credit Marci Heiser

to the summer solstice the bees are noticeably gentler than after the summer solstice. I would consider opening a hive with no smoke in spring if I knew I would only be there a few seconds. But that is not advisable in fall. For example, I popped open the Braden girls just long enough to put on a candy super and candy. The hive was open maybe 20 seconds. It was January and I got 6 stings. I had my veil on. We could do a similar move in spring, during the flow, with no stings.

I open up my colonies as little as possible. Each one is a superorganism. To open up the hive is nearly equivalent to surgery on a human. That being said, it is not my recommendation that beekeeping students open their hives as little as possible. As a new beekeeper you will gain valuable insight and knowledge every time you look. You should be looking at your bees frequently until you are familiar enough with what is happening in the hive that you can predict it without looking. Take notes about what you see. Yes, this will negatively impact your honey production and it will, potentially, negatively impact your colony health, but the knowledge gained in your early years is worth this cost.

Your beekeeper's suit should be dirty in your first year. If you get to summer solstice and it doesn't need to be washed, you are not looking at your bees enough.

Because I want to open my colonies as little as possible, I combine efforts when I do need to open. During "first inspection" we can do several chores, more than any other time of year.

1. Remove any uneaten candy.
2. Remove the candy super.
3. Remove excess honey (spring harvest).
4. Assess queen health.
5. Assess colony strength and make notes of how many splits can be done with this colony.
6. Scrape burr comb.
7. Clean up and perhaps replace bottom board (replace when colony is suffering from nosema).
8. Prevent swarms using checkerboarding technique.
9. Remove old comb (finished its third or fourth year) from bottom deep (this is Langstroth only).
10. Eliminate old or damaged equipment.

This is the smallest population of bees we will see all year. We take full advantage of that and do as much work as possible. We must show up with extra equipment (frames, bottom boards, deeps) in tow just in case we need it.

Tools of the Trade

© steve kennedy 2014

Every profession and every trade is made to look quite easily done, by the experts. It would not be unusual to think, "That looks easy, I'm sure I could do that!" But when you try it, you find out that there is much depth of knowledge and subtle skills not easily noticed by the casual observer. Observers cannot account for what's going through the head of the expert. The expert is likely considering a host of factors based on years of experience which may alter how she moves and what she chooses to do.

Meanwhile, you have to start somewhere. Getting started with a good set of necessary tools will help you onto the right path. Beekeeping is ripe with gadgets and doodads you can purchase and enjoy if you want. This chapter will stick to very minimal tools. Most will be necessary. A few are optional, but nice to have.

Illustration 50: Different suit preferences. Photo credit: Don Studinski

Bee suit and variations

Honeybee instinct to sting is primarily triggered by their need to defend "home." Therefore, the most likely person to be stung by a honeybee is the beekeeper, the person opening and poking around in that home. The bee suit you pick is an issue of personal preference. However, you must have a means of protection. It is not realistic to be a beekeeper and not get stung. This is true even with a bee suit on. Honeybees will sometimes sting through a layer of fabric. The suit will make a huge difference in how many times you get stung, as will effective use of your smoker.

Illustration 51: Jacket. Cheap pants. Boots. Photo credit: Joel Studinski

One of the most important parts of your suit is the veil. Suits come with a variety of veils to protect your head and face. Some are in combination with a hard hat, called a **helmet**. The person to the left in in the photo above is using a helmet with a full-body suit. Another veil type does not have the hard hat. The veil is usually part of a jacket and it's called a **hood**. The hood comes with plastic reinforcement to keep the netting away from your face. The folks to the right are wearing hoods.

Some suits are just a jacket, others are a full body suit. Should you choose just the jacket, keep in mind that you will probably still want a second layer for your legs. Bees can and do sting through a single layer of cloth. My choice has been a cheap pair of dress slacks from the second-hand store. I buy them extra large to go over my jeans. In any case, you want light colors. Bees seem more inclined to sting dark colors.

My hooded (dome) jacket is made by Dadant. I have enjoyed it for several years, however, I don't always like the way the dome touches my nose. Sometimes, all the bees need is that second of access to my nose and I've got a sting. My son likes his suit which came with an attachable clear view hat. Inspired by that, when it came time to get a separate veil, I chose the clear view type.

At a minimum, pick a veil and a jacket that you are comfortable using.

Illustration 52: Clear view veil alone with tie down. Photo credit: Bobbi Storrs

One handy "extra" for your wardrobe is a helmet and veil or a clear view veil with just straps for your arm pits. This is a great item for the heat of summer when you are willing to take a sting or two in order to avoid wearing the jacket or full bee suit. Heat stroke is real and you need to be careful during the hottest part of the year. The inconvenient scenario you want to avoid goes something like this:

You are inspecting in full sun late in July (it is <u>hot</u>). The hive is open and disassembled. You are inspecting frames. Bees are in the air and hive boxes are spread out all around you. Right about then, you start to feel dizzy and have chills. These are signs of heat stroke. You know you are in trouble and you need to get out of that bee suit to cool down. Alas, shedding clothes now means guaranteed stinging. But it's going to take some time to put everything back together, even if you are willing to smash a bunch of bees in the process. And smashing bees increases the number interested in stinging. Hopefully, that's enough of a description to motivate you to avoid getting over heated. Of course, I'm just making this scenario up. I have never done this myself. Or have I? I can't remember. Must be brain damage from heat stroke.

Gloves

The gloves you buy at the beekeeper store are usually made of soft cowhide or goat leather with a vented sleeve. Other options include canvas and plastic coated. Another much less expensive option is kitchen cleaning gloves or even garden work gloves if the sleeve part is not important to you. Be sure you are able to close off the opening around your wrist or arm. Bees will crawl in and sting if you allow them an opening.

Washing

Besides the obvious dirt, honey, wax and propolis you will get on your suit and gloves, you will also collect venom. As this builds up, it will be more likely to trigger aggressive behavior from the guards. You want to avoid this. Wash your outfit regularly, though washing is probably not necessary every trip. Don't be fooled into using the washing machine. Yes, I really did that! The veil and glove vents were torn badly. Use hand washing. Also, consider using fragrance free biodegradable soap. Fragrance free because you

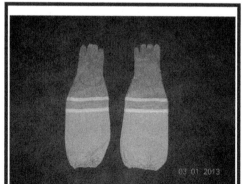

Illustration 53: Traditional beekeeper's gloves including vents and goat skin hands. Photo credit: Don Studinski

never know what smell is going to set them off. Biodegradable because you want all your activities to be earth friendly, for the bees.

Treating a Sting

You would think I could list a set of remedies (tools) that beekeepers could use for treating a sting. Scientifically, this turns out to be false. Honeybee stings are somewhat uncomfortable, but generally not life threatening. Depending on the reference, somewhere between 1 and 2% of adults (less than 0.5% of children) show serious, anaphylactic shock due to bee venom. Beekeepers having repeated stings over time can develop these symptoms, or potentially build immunity. Stings may be painful for a few minutes to a few hours. Localized fever is common in the swollen area. Swelling and itching may persist for several days. Scratching will increase swelling and itching. Should symptoms persist more than a week, seek medical advice.

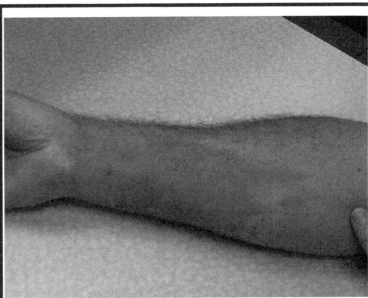

Illustration 54: Honeybee sting on wrist. Not an allergic reaction. Red is a normal reaction. Photo credit: Don Studinski

On the outside chance that, immediately after a sting, you experience a pounding heart, dizziness, nausea or difficulty breathing, you should seek medical help immediately. This can be life threatening.

There are numerous folk remedies which may provide some symptom relief due to placebo effect. Actually, cold compress can help with swelling and pain, but that's all there is. Various itching suppression remedies can help with itching.

Foot Protection

I really like my high-top leather boots with steel toes. They prevent bees from getting to my ankles and give me a nice long tie string which allows me to include my pant legs inside the tie. If you forget to tie your pant legs like that, then expect a visitor up your leg, perhaps at an inconvenient time like when you have just left the apiary and are driving away at highway speed. Again, not that I would know! This type of boot is expensive and is clearly not a necessary thing, but it sure helps. You should focus on the foot ware attributes: cover feet and ankles thoroughly. Make sure you can stay focused on being calm as you work the bees. Many stings will not help you be calm.

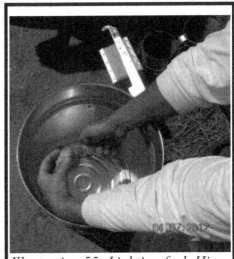

Illustration 55: Lighting fuel. Hive tool. Picture credit: Marci Heiser

Illustration 56: Smoker getting started. Photo credit: Don Studinski

Hive tools

There are probably hundreds of hive tools from which to choose. If you want to save money, you probably have something that could be used perfectly well already in your toolbox. It needs to be able to pry frame boxes apart and able to pry a propolis-stuck frame up out of the box. It's handy if it can also separate one frame from the next. If money is not so important, you may want to try two or three to see how you like them. This tool is a must. I recommend having at least two of the one you choose as your preferred hive tool. This is because if you drop one in the grass in the middle of an inspection and cannot find it, you want to have a backup. One hive tool is pictured just above the can in the "Lighting **fuel**" picture on the previous page. It's the tool sitting on the smoker brightly reflecting the sun.

Smoker

I encourage you to use a smoker. This makes a significant difference in how defensive the bees are. When we smoke them, it masks the "attack" pheromone that the guards may emit (it smells like bananas). Smoke also causes the workers to gorge themselves on honey in anticipation of making an emergency exit due to fire. Once full of honey, the workers are a bit sluggish and less likely to get excited about attacking.

Lighting a smoker can take a while. You should be carrying a lighter and a backup lighter. My smoker can is a handy wind break for lighting my smoker fuel. It's not unusual for it to take longer to light the smoker than it takes to do the inspection. That is, if you

Illustration 57: Working bees without smoke. Photo credit: Marci Heiser

don't have a propane torch. With a propane torch, you can fill the smoker with fuel and blow the flame on the outside toward the bottom. You are heating the smoker itself, but lighting the fuel inside. This is a very quick way to light your smoker. It would be convenient to be able to bring the smoker, still lit, from one apiary to the next. To do this, I use a small metal trash can. It just fits the smoker. This was found at a pet store. I think it is intended for pet food. It's roughly the size of a 5 gallon bucket. It works great and allows me to place the smoker, hot, into the can, put on the lid and place the whole thing into my car as I head to the next location. Lighters and a smoker are essential. I suppose the smoker can is optional, but I wouldn't be without it.

You can use almost anything for fuel, but you want to get cool white smoke and you want to avoid accidentally spraying the bees with embers. You also want it to stay lit for the duration of your apiary visit. I prefer to purchase smoker fuel made from compressed cotton fiber. It's called "Smoker Fiber Fuel." It's easy to light, stays lit, and when you've used it for a while you know about the right size to break off for any given visit based on how long you plan to stay. I also supplement my purchased fuel with stuff from home. Dryer lint is almost exactly like the compressed cotton, just not as compressed, and dried citrus peels work dandy.

Student Questions

Linda writes: I need to practice with my smoker a lot more. I ended up inspecting without a smoker because I couldn't keep it lit. Any tips? Probably didn't help that it was super windy.

Answer: I too have had many experiences inspecting smokeless, unintentionally. Just when I think it's lit I open up the hive and get in the middle of what I need to do, next thing you know, it's out. My suggestion is to use that cotton fiber smoker fuel. Once it's lit properly, the wind will help keep it lit, but getting it lit is the key. If you feel like an additional investment, a propane torch is very helpful for

Illustration 58: Harvest using smoke. No stings. Picture credit: Susan Sommers

lighting the fuel quickly. Alas, this too takes some practice to use. If you tip it too much, it will go out. Another trick is to keep opening the top when you set the smoker down. Fire needs oxygen. If you leave it sitting with the top on, the fire gets oxygen starved and will go out.

Attitude

A surprising, but powerful tool you will quickly learn to use is your attitude. If you approach the bees feeling agitated, they will tell you they don't really like you hanging around. If you approach the bees feeling anxious, you will get the same response. Honeybees are an amazing biofeedback mechanism. You will quickly learn to calm yourself and you will get the response you want.

Happily, this skill will spill over into the rest of your life. Many beekeepers say they are much more

Illustration 59: Langstroth as student prepares for split. Photo credit: Don Studinski

relaxed since they have been working with bees. There are many factors at play, for example, just being outdoors, connecting with nature, is calming and nearly everything in the hive is of medicinal value, including the venom. I frequently pop propolis into my mouth while performing chores at the hive. But, don't ignore the power of communication between the colony and the beekeeper. They want you to be calm and you, in return, want them to be calm. Everyone is sending this message: be calm. And it works! If, on the other hand, you smell an odor like bananas, then the "calm" message has failed. That might be a good time to close up and walk away. That banana smell is bee speak for "defend."

Hive Types

Every hive type has both benefits and draw backs.

Langstroth:

Langstroth hives have been around and successful for over 100 years. Dimensions are standardized which makes mixing and matching equipment relatively easy, however, you will run into slight differences from vendor to vendor which can be inconvenient. The reason mixing is important is because there will be frequent times you want to move a frame or a whole box from one colony to another either to help with their food needs or to accomplish a split or a merge. Another benefit is that many of the tools on the market were specifically designed for Langstroth. This includes, grips, hive tools, **extractor**s and more. In fact, if someone uses the term "hive" without being more specific, it is a safe bet that they are referring to Langstroth equipment. Screened bottom boards, which some prefer, are readily available.

The Langstroth draw backs include a relatively high cost. Many of the frames include the plastic foundation which is not easily accepted by the bees. Frames with solid foundation require the bees to maneuver around an extreme end to move from frame to frame. This can be a disadvantage for the cold weather cluster.

Top-bar hive (TBH):

The primary motivation for TBH has, historically, been cost. These designs are so simple, you can quite literally build one with just a little scrap lumber. The bees build their own "frame" comb, allowing the keeper to "see" what the bees choose somewhat more naturally. You can easily standardize your chosen dimensions which facilitates mixing and matching equipment from hive to hive. The hives are typically more horizontal than vertical which can afford some benefit concerning wind, but since most folks typically put them on a stand of some sort, this benefit is debatable. The way the top bars lay out horizontally allows the beekeeper to disturb small portions of the hive at a time. This is

Illustration 60: Top-bar hive. Photo credit: Don Studinski

helpful for keeping the bees and the beekeeper calm. Many of the designs include a window which allows keepers to see what the girls are doing. Freshly caught swarms drop in very easily and they quickly draw comb which is fun to watch. **Abscond**ing is rare.

The down side is primarily the complete incompatibility with Langstroth equipment. Anything you want to move from Lang to TBH or visa-verse requires special, usually hand-built, equipment. Further, you cannot count on compatibility from TBH to TBH unless you built them all using the same design. Bill's is different from mine and mine is different from Valerie's. Make no assumptions! Another strong down side is that there are no foundations.

Although I've heard of keepers spinning these frames to get honey, they are fragile and breakage is common. Honey extraction will typically involve **crushed comb**, gravity and a **strainer**. It's more messy than Lang, but any honey extraction is messy. Most designs do not accommodate standard feeders. Even feeding candy can be somewhat inconvenient if the brood is far from the candy space.

Warré:

This vertical hive with frames is somewhat similar to a Langstroth and is sometimes referred to as a vertical top-bar hive. The frames vary: Just top bars like a TBH, top bars with the downward sides, or a full rectangular frame, and further, each of these may include a wax building edge to

Illustration 61: Warré parts left to right: roof which fits over the quilt, quilt (for insulation material, like wood chips) which goes on the top deep, two deeps on a bottom board with legs. Photo credit: David Braden

encourage the bees to "organize," or no wax building edge. Another difference from Lang is that these include an insulating layer called a "**quilt**" above the frames, but immediately below the roof. Like TBH, these come in many size variations; there seems to be no standard, unless you build your own consistently. The entrance is unique and smaller than Lang. Management is also unique in that new boxes are added at the bottom, raising the stack. Honey is harvested from the top. This provides an automatic rotation of wax out of the hive which is what I find most attractive about this hive design. Keepers will tend to rotate wax out before significant pesticide buildup can happen. Warré management also emphasizes disturbing the bees as little as possible.

Negatives about Warré would include that "insert at the bottom" management because it's potentially heavy to lift the boxes above without breaking the seal between boxes and Warré keepers usually prefer to avoid breaking that seal whenever possible. Another negative could be the relatively thin design of the boxes, compared to Lang, which may make them more vulnerable to wind. Of course, with no foundation, the honey frames will be difficult to spin using a standard extractor, but apparently, it can be done. Keepers would be wise to expect a crushed comb honey harvest and plan accordingly. 2012 is my first year keeping Warré-style hives. Ours are a modified design created by David Braden. All the details are on my website.[23] I will expand on these if my experience is as favorable as I expect.

---

23 http://honeybeekeep.com/content/building-7-modified-warre-bee-hives, visited 3/13/2014

April … Swarms, Nucs and Packages

© steve kennedy 2014

Spring is busy for beekeepers. Swarms can come any time. You are basically on-call for the entire season. This will not slow up until summer. Some of us will be performing splits which can involve making nucs. Both swarms and nucs are all about the honeybee instinct to multiply. It's a good thing but very demanding all at once. Packages are also delivered this time of year. Packages may need feeding and definitely need regular inspection. In any case, it's all more demand on the beekeeper's time. Your supers should be on or going on very soon. Now is the time of abundant nectar flow. If you are going to get a harvest this year, you will likely know it very soon.

Inverting Deeps

Some **beeks** (short for beekeepers) say that inverting deeps in early spring is a good swarm prevention technique. I do not agree because this logic does not account for all the factors needing consideration. Here is what is meant by inverting deeps. In many cases, a colony of bees living in Langstroth equipment is given two deeps for their brood area. The theory goes that the bees will have moved up into the top deep during winter as they consume their stores. In many cases, this is correct. This would leave an empty deep under the bees. If you have verified that the bees are living, entirely, in the upper deep, then you can invert without difficulty. This may or may not prevent a swarm. The whole brood nest is in the upper deep. You move it to the lower deep position. The lower deep was empty. It is moved to the upper deep position. In this scenario, you can invert deeps without causing a big disruption to the bees.

This line of thinking also assumes that the bees prefer to build "up" in general. Thus, it is reasoned, if we swap the deep positions in early spring, moving that empty deep above the bees and the full deep to the bottom position, we give the bees room to again start their "natural" upward building as they expand their brood nest. Given plenty of room for expansion above the nest, the bees will choose not to swarm.

But what if the bees are still living in both deeps?

Linda has been getting frustrated lately. She inverted her deeps. Oops. She writes:

> After the third bee sting this week, while NOT working with the bees, I started wondering about my future as a beekeeper. My angry bees are the result of reversing hive bodies in my Langstroth - a very disruptive action. They seem to remember that it was me that did that to them. They know me and they are holding a grudge.
>
> My only hope is that my current bees "work themselves to death" in the coming weeks and future generations will know nothing of the hive body

Illustration 62: Langstroth and Warré. Photo credit: Linda Chumbley

inversion. Hopefully, they will not pass along my information in the weekly staff meetings!

What Linda did to the Lang may have disorganized their brood nest, or worse, accidentally killed their queen.

Consider this scenario: Picture a volleyball inside that hive. That's the shape of their nest. The top part of the ball is in the upper deep. The bottom of the ball is in the lower deep. Honey is above the top of the ball. Below the bottom of the ball is probably empty. If this is the case, then Linda took the bottom of the ball and moved it to the top box position. That puts brood at the top of the hive and the lower empty part in the middle of the hive. She moved the top of the ball to the bottom box position. Brood is near the bottom board and honey is in the middle where the girls don't normally put it. The nest has been split into two pieces; the bottom part of the ball is at the top and the top part of the ball is at the bottom. They have to rearrange everything. Also, they may lose some brood if they cannot keep the split nest warm. Such a condition can make for angry bees and, yes, they know you.

Also, honeybees are frequently "angry" when they are queenless. If the queen has been accidentally killed during a manipulation, the beekeeper may notice the anger until they are queen right again. When going through an angry spell, try to find something to block the guard bees' vision. This will allow you to move around the yard without them knowing you are there. I'm using hay bales for this.

Illustration 63: Van Gordon colony, three deeps plus three supers. Photo credit: Don Studinski

Rather than the two deep configuration Linda was using, I prefer to use three deeps. I find that during the warm months, the bees will expand their brood nest to take advantage of all three deeps. I want them to have plenty of room for honey above their nest before reaching the supers where my honey will be stored. This helps prevent the queen from entering the honey supers and allows me to skip using a queen excluder.

Furthermore, bees naturally build down, not up. Consider a tree cavity hive. They start at the top and begin building down. As they need to expand, the uppermost comb becomes strictly honey storage. Below honey is always the arc of bee bread and below that is brood nest. At the bottom will be drone brood and potentially queen cells for swarming. Comb to the far sides may also be honey storage, with bee bread inside that, followed by brood in the middle. This is always the configuration honeybees prefer. Yes, they will have moved up during winter as they consume their stores, but there is no need to invert. They will naturally move the brood nest down as the season progresses.

This is why Emile Warré suggests that new boxes be added at the bottom, the nadir.

> Will lifting the Warré boxes to add another hive body create a similar "angry bee" scenario? And, what about the weight of the Warré boxes as the bees fill with brood and honey?

When we add a Warré box to the bottom of the stack, we do take apart the other boxes to get at the bottom board, however, we will put it all back the way it was on top of the new box. It should be significantly less disruptive. As the boxes fill with honey, they will become heavy, so careful lifting will be necessary to prevent back injury.

Acknowledge Beekeeping Challenges

These are troubled times for our industry. We have experienced 30-40% losses annually since 2006. As of this writing, March of 2013, I have a 36% loss for the winter of 2012/2013, that's 9 of 25 dead.

You should approach involvement in such an hobby carefully. Get your eyes open to the potential losses. This may or may not be for you.

Hurtles we face:

1. pesticides
2. government agencies that have lost their way
3. parasites, like mites
4. diseases

Illustration 64: Cherokee queen cell early spring. Photo credit: Danielle Bryan

These are just the "out of the ordinary" challenges of our current time. There are also all the typical challenges that beekeepers have faced for millennia.

It is not at all clear to me that these obstacles can be overcome. I am particularly concerned about the systemic poisons, like neonicotinoids, and their multi-year half-life in the soil. I'm sticking with beekeeping because it is clear to me that, without the bees, we have even greater difficulties. It's not all fun, but there certainly is fun along the way.

The Split Process

Splitting a colony is a process requiring three trips to the apiary. There are multiple ways to approach the splitting process. I will describe two here.

Prerequisite: a healthy colony with three or more frames of brood and eight or more frames of bees.

A split is essentially an artificial swarm which satisfies the swarm instinct and places the resulting new colony(s) exactly where you want them. It's a powerful way to avoid swarms, multiply your bees and control where your bees live. Your neighbors will benefit from it as much as you do, even if they don't know it.

We begin this discussion describing a method that includes moving the existing queen.

First trip to the apiary:

Open the hive and search, frame by frame, until you find the queen. Move the queen, along with brood, bee bread, honey and plenty of bees to a separate hive or nuc. This new hive must be moved at least two miles away to prevent the foragers from drifting back to the "home" from which they came, the original hive. Having moved two miles, foragers will reorient to their new location.

Leave, in the original hive, the remaining two or more frames of brood, bee bread, frames of honey and plenty of nurse bees covering the brood. These brood frames must have eggs or larvae younger than days four, five or six. Only a young larva or egg is a candidate to

Illustration 65: Birch split preparaton. Photo credit: Don Studinski

become a queen. A larva less than six days old is as small or <u>smaller</u> than 1/6th of the bottom of a worker cell. If you fail to leave viable candidates, then you have just made the bees hopelessly queenless. My eyes cannot see an egg. I can barely see a larva small enough to become a queen. Therefore, I let the girls pick their own new queen. If your eyes are good, then you can "encourage" or "select" the cell(s) to become a queen by crushing the bottom cell wall to the mid rib. You must do this without damaging the egg or larva in the cell. If you get good at this, then you can actually determine exactly how many queen cells the girls will make. This is a powerful concept found by Mel Disselkeon. I highly recommend his website: mdasplitter.com.

Close up the original, now queenless, hive. The bees will know they are queenless within four hours. They will consider this to be an emergency. They will immediately begin new queen preparations. The continued existence of the colony depends on it, and they know this. They will be performing what is called an emergency supersedure. Supersedure is the term used when a colony replaces their queen. This is different from swarm preparation. Being queenless, these girls will be testy. When you come back, you should expect a noticeably more hostile attitude. They don't like being queenless. The next couple of weeks are the most likely time to have casual passers-by get stung by your bees. Consider putting up a visual "screen" so the guards cannot easily see human traffic in front of the hive. I've used hay bales for this purpose. It helps keep angry bees off the gardeners.

Second trip to the apiary:

Approximately one week later, return to the original, now queenless, hive. The girls will have made significant progress producing queen cells. Be careful during this inspection. Queen cells will be sticking out into the space between frames. You must not break them or damage them in any way. Avoid bumps, they are made of soft wax and can be easily damaged. A queen cell is capped on the 9th day after the egg was laid. Therefore, when you find the capped cells will depend on the age of the eggs and larvae chosen for queen cell production. What you hope to find is two or more queen cells on each brood frame left behind.

In this case we are making two new colonies from the one we left queenless. Move at least one brood frame, bee bread, honey frames and bees to each hive. You want at least two queen cells in each hive, but probably not more than four queen cells. One will become your new queen. The others are sacrificed. The first queen to emerge will kill her competitor(s). Should two emerge at the same time, then they will fight to the death. Killing extra queens, in any case, takes energy and is dangerous. Therefore, you don't want more than four queen cells in each hive. If you have extra, destroy them as you separate the frames. Generally, bigger queen cells are superior queens.

Illustration 66: Birch split. Healthy colony. Photo credit: Don Studinski

At this point we have two hives of sisters. They will not fight one another. If one or more queens fail, then we can combine these colonies right back together without difficulty. Position the hives together as closely as possible. Drifting foragers is not a problem at this stage. In fact, it's a good thing if it evens out the incoming nectar and pollen. Close up the hives and leave them for three or four weeks.

Third trip to the apiary:

The queens complete their metamorphosis and emerge on the 16th day after the egg was laid; that's one week after she was capped. They will spend about a week maturing; that's two weeks. Then, these virgins will take their nuptial flights, getting serviced by 10 to 20 drones. More promiscuity creates more diversity which is beneficial for nature in selecting genes that work. Mating may also take about a week, that's three weeks so far. Once compete, she will begin to lay. If you can see eggs, you can make your third trip as soon as she begins to lay. Again, I cannot see eggs, therefore, I need to wait about one more week before I inspect, that's four weeks. When I look inside again, what I want to see is a nice laying pattern with plenty of white larvae, which, by that time, I can see. I'll probably see capped brood as well because they will begin putting on caps nine days after she begins to lay.

Assuming I find what I want, the split is complete and successful. Hives may be relocated as desired. Should one or more of the splits fail, then I'm glad they are all still positioned right together because the next step is to combine the sisters back together. Of course, another alternative is to purchase a queen. The worst case would be that none of the new queens is successfully laying and all the splits must be recombined with the original queen (which is now two miles away). In that case, I would move the original queen back to where she started.

Now we will cover a method that does not involve finding and moving the existing queen. This is my preferred approach because by avoiding the need to find the queen, we disturb the bees less.

Illustration 67: Birch queen cell check. Photo credit: Don Studinski

Most of this method is exactly like the one where you find the queen; so, I don't plan to repeat most of the instructions. In this method, we don't bother finding the queen and we don't move any hive two or more miles away. In permaculture, we try to find ways to work with nature such that we do less work. In this method, you can have the benefit of less work. The trade off is that you may get fewer colonies in the end.

Simply split the brood, bee bread, honey and bees evenly across as many hives as you can safely do given the strength of the original colony. This time, you don't know where the queen is and you don't really care. However, you do know that one of the hives has a queen and that this will be a powerful draw for the foragers. We want the foragers to be confused such that foragers return to each of the hives; keeping the bee population up in each. Therefore, we want each hive to be close to, but not exactly where, the original hive was located. This has to do with entrance position. Force the foragers to "return" to a variety of entrances, not knowing which one will contain the queen.

This is easiest to do if you are only splitting from one hive to two. In this case, you place both hives on the hive stand such that half of their entrance is covering the space where the original entrance was. In this manner, the foragers will sometimes enter the hive on the left and sometimes enter the hive on the right. Eventually, both with be queen right again.

If you have more than two, then I suggest not leaving any on that original hive position. Make the bees go and find an entrance. Hopefully, they will even out over all the new hives. Place the entrances close, but not exactly where the original entrance was.

This method is especially attractive when you can begin splitting on a box basis rather than a frame basis. When your colonies get particularly strong, you can move to a configuration that uses three brood deeps rather than two. In this case, when the queen is laying across a brood nest that spans two of the three deeps, then you can split using whole boxes. It's very quick and efficient. Grab a deep with brood and move it to a new bottom board. It's a whole deep full of bees. Nurse bees will not leave their brood. Some foragers will drift home, but other bees will become new foragers and return to their new location. This is the least disruptive split for a colony. The entire deep is left undisturbed with brood, bee bread, honey and bees. The other two deeps, which have been left behind, are also minimally disturbed. This aids all the colonies to quickly recover from the split process.

A week later, return to discover which hive is queenless and which contained the existing queen. The hive where you find the queen cells is the one you had rendered queenless. In any case, you are again hoping to find multiple queen cells on multiple frames. Assess what you have and separate boxes and frames as described above until you have created all the new colonies you wish or can safely create.

Splits Performed

Today we split the Birch Street girls and Mo's Pink girls. We also helped Mo with some hive configuration adjustments. All the girls seemed quite happy with the exception of three guards which followed us to the car at Mo's house. However, they did not sting.

The Birch Street girls started in a three deep configuration. They were very strong indeed. We found beautiful brood in the upper deep with plenty of honey and bee bread. This allowed us to do a box split. We shifted the existing hive about a half-hive to the west. I set up a new bottom board just on the east side of it. We then set an empty deep on the bottom board. We loaded that bottom deep with drawn comb that I had retrieved from my shed.

Illustration 68: Queen cell found in progress. What does that tell you about the cell chosen to be the new queen? Photo credit: Don Studinski

There was some damage to the comb by wax moths, however, not bad enough to make the comb unusable. Then, we were able to take the entire top deep from the top of the stack and set it on the deep we had just loaded with comb. This gave the most easterly hive the entire "upper" portion of the basketball sized brood nest and left the most westerly hive with the entire "lower" portion of the brood nest.

That sets up both hives with eggs, larvae, honey, bee bread and honeybees. By now, all the girls know who is queen right and who is queenless. They know within four hours if they are queenless. Those that are queenless are in an "emergency" state. They must make a new queen or perish. They are beginning work on this even now.

At Mo's house, we didn't have the equipment for a box split, so we did a frame split with three frames of brood moved to the new hive along with honey and bee bread. They will be a very small bunch for a while, but they should make it. We are hoping for good weather or at least to avoid any plunges down into single digits.

Next week, we go back in to see who is making queen cells and how many. If there are more than four queen cells, then we must kill the extra. We only want one, plus a few extra for insurance. Should more than one emerge, they will fight to the death. More likely, the first to emerge will find the others and kill them before they can emerge.

Queen Cells

Illustration 69: Open queen cells in progress. Photo credit: Don Studinski

When you see a queen cell, it can give you insight about what's happening in the hive. You will want to take note of position and number of cells. You may have been trying for queens, or the cells may be an unpleasant surprise. Obviously, when trying for new queens, a minimum of one queen cell is required. An extra queen cell is comforting insurance. The bees clearly don't want to put all their hope into one cell, that's why they build extras. But, extras can be taken to an extreme. I suggest you not allow more than four to stay for the duration of the cycle in any one hive.

Can we find a way to use extra queen cells to further expand our bee population instead of smashing them? Of course we can. An alternative to death is to lift it and offer it to a fellow beekeeper that may be in need.

We are likely to have extra queen cells while performing our split process. Consider the two splits we have underway.

Birch Street girls: We will inspect for queen cells on Saturday. These colonies are particularly strong in both hives. We could, potentially, split them into a third or even a fourth. I stopped by the Grange apiary recently and looked at the split hives. What I saw was very encouraging. A great deal of flying honeybees were at each front entrance. There's a good chance we will find multiple queen cells in the hive that is generating a new queen. We MUST have at least one queen cell for there to be any chance of a successful split. It's wise to leave a second, third or fourth as insurance, in case the first doesn't emerge. Beyond two, I might kill. I would certainly kill beyond four to avoid triggering an unwanted swarm.

Mo's Pink girls: This is another good strong colony that might make more than two queen cells. This colony cannot support a split to a third or a fourth, but could, potentially, supply a frame with a queen cell for another colony. Mo looked at the split hives and said there was good activity at both. This is also very encouraging.

The target for any extra queen cells will, necessarily, be at the Grange apiary (Birch Street colonies). But the source could be either the Grange or Mo's house. The trick is moving the cells while keeping them warm enough to remain viable.

The weather started turning bad. Therefore, to take advantage of weather right for the occasion, we inspected our queen cells on 4/13/2012, a bit earlier than planned. All was well in the Birch hives. Mama Birch was happily in our east hive with the new bottom deep. The existing two deep hive on the west had produced an abundance of queen cells. Therefore, we didn't need to move queen cells from Mo's Pink girls to the Grange apiary. We killed all beyond the four biggest ones (for each new colony) and split the two deep hive again. This created two daughter colonies plus our original, Mama Birch. A very happy checkpoint. Now, the queens must emerge, fly, mate and return to successfully lay. All this must be confirmed before the split is complete.

Meanwhile, clients have many questions this time of year, mostly about queen cells and swarming. One second year beekeeper found eight queen cells in a TBH. "What should I do now?" they ask. "I only have the one hive in my backyard, I live in a suburb and I'm not equipped to do a split. My hive faces the street behind my house so they probably won't swarm straight into a neighbor's attic but I'd like to keep everyone happy."

In this case, the beekeeper has probably waited too long to be able to stop the swarm. In fact, they may swarm more than once. My advice is to leave the 4 biggest cells and kill the rest. The swarm should show up on the day the cells are capped. Watch carefully and you may catch your own swarm.

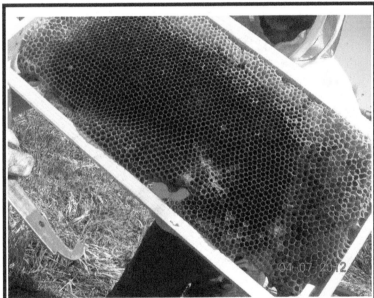

Illustration 70: Minor wax moth damage. Photo credit: Chris McCune

This brings up some interesting points about queen cells. They appear in different positions depending on the motivation of the bees in building the cell.

If you see just queen cups with nothing in them, there is no reason to panic. There are usually some of these in any hive. Seeing cups is not the same as seeing queen cells.

Queen cells near the bottom of the frames are usually an indication that your bees have decided to swarm and there is little you can do to stop it. Even if you kill all the queen cells, they will build more. You needed to free up space for them about a month earlier. If this is your case, you need to find a friend with empty hive equipment that will allow you to split immediately, as in today. This would allow you to place a new colony in a new hive in a controlled manner. They may still swarm, but it's worth a shot.

If queen cells toward the edge of the frame are already capped, then chances are the swarm has already left. In this case, leave the hive as is for three weeks. Do not kill those queen cells unless you are sure you have found the old queen. The three weeks will allow the new queen to emerge, sexually mature and fly for mating. Before that three weeks is complete, the bees can be funny about a new queen and **ball** her (kill her) if they are disturbed. When she completes her mating, she will begin to lay. If all is well, and you are capable of seeing eggs, you can inspect after the three weeks. If you are not capable of seeing eggs, like me, you have to wait four weeks to see the big larvae and capped brood.

When the colony collectively agrees to multiply, that is, create a swarm, they prepare their swarm cells in the preferred position, on an edge. They have the luxury of time and cooperation from the queen. When the cup is ready, the queen will lay a fertile egg and the nurse bees can raise the virgin queen. Compare this to when the colony unexpectedly finds itself with a failing, missing or dead queen. In this case, the colony is in an emergency. They must make use of whatever larvae they have which are queen candidates. They do not have the luxury of time and they have no queen to help by producing a new fertile egg. Therefore, you get the queen cells mid-frame, or where ever the candidate larvae were at that time.

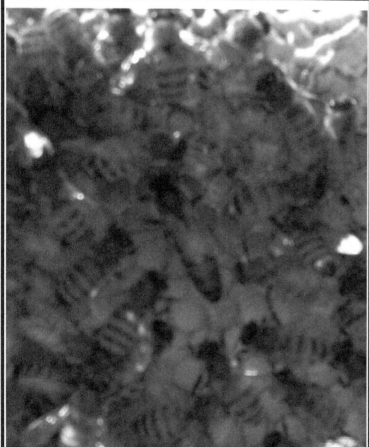

Illustration 71: Her royal highness, queen Mama Birch. Photo credit: Don Studinski

If the queen cells are mid frame, then your bees may have decided to replace their failing queen. This is called supersedure. They may be queenless or she may be failing. These cells should be left undisturbed unless there are more than four. The first queen to emerge will probably kill the other queens still in cell. If this doesn't happen, queens can fight to the death, or one of the queens may be run off and a swarm may issue. If you kill all but a few, you may be able to avoid a swarm and be queen right.

Split Progress

The queen was in the hive to the east and I got her picture. Plenty of cells in the hive to the west. So, I divided the two boxes from the hive on the west and gave each about four queen cells. The queen and the two boxes on the east got moved to Golden, 20 miles south.

Mo's ladies were also progressing right along with one queen cell capped and several more in progress. I left it as is. The girls will pick the queen they want. Mo's new hive was a dead-out with left over honey. The hive had been invaded by a mouse. It experienced significant damage on the 3rd and 4th frames of the top box. We washed these with the hose to get the feces and urine off. None of the honey is for human consumption, but it will be fine for feeding bees.

It's a great day. One strong hive has become three. Another has become two. I'm very pleased. Now we wait, no peeking. One week before the queens emerge. Two weeks for her to get sexually mature and to take nuptial flights. One more week for eggs laid to become visible larvae. Then we peek. That will be 5/12. Obviously, much can happen out in the dangers of nature, so we will just have to wait and see.

Just as important as "no peeking" is the check for a healthy queen afterward. Her ability to fly and mate can be negatively affected by weather. You can end up with a queen that is alive, but useless. You must check the laying pattern. You want a nice tight pattern that fills a frame. It could be a football shape or it could be wall to wall brood. If she is not performing, you will want to consider a replacement. It doesn't pay to wait on this decision. You need a healthy, laying queen right now. Don't wait on a second rate queen to get better.

Unplanned Split

If you don't stay ahead of the swarming instinct, you can have a situation where you need to perform a split "right now" as is Valerie's case. These girls need a third deep and a honey super right away. They were boiling over, clearly on the verge of swarming. We got together on 4/21 to perform the split in an effort to satisfy their need to swarm. This was a case where we had to find the queen because Valerie was keeping the mama and sending me with the queen cells. Find one bee among about 50,000. It's not my favorite chore. We searched all 20 frames, no luck. Searched 19 frames a second time, no luck. Picked up the last frame with me looking at one side and Valerie looking at the other. Valerie says,

Illustration 72: Boiling over. Photo credit: Linda Chumbley

"maybe." I said, "Don't take your eyes off her." I slowly turned the frame so I could see and, sure enough, there she was. We had separated all the queen cells, about four, into a different deep. We replaced queen "Liz" in the original hive and took the three brood frames and two honey frames for the split. These ladies were immediately transported to their new home.

Illustration 73: Nuc, short for nucleus hive. Photo credit: Ruth Rinehart

Swarm Season Kickoff

First bloom was a full month earlier than normal in 2012. Specifically, first bloom in 2012 was 3/14 whereas first bloom in 2011 was 4/9. With the trees and flowers blooming early, it is possible that swarm season could start early. We do not know exactly what to expect. Normally, I don't expect the first swarm until mid April. This year, who knows?

Illustration 74: Nuc vent close up. Photo credit: Ruth Rinehart

It is best that we prepare well in advance. That means now is the time to get your gear ready for swarms. These items should be in your vehicle in case you need to roll on short notice: bee suit, veil, gloves, hive tool, smoker and a nuc or equivalent. This might be one of: cardboard box with air holes and tape, some equivalent box with ventilation or a real nuc.

Never put a colony into an unventilated container to transport – the bees (generating much heat) will over heat and die!

Illustration 75: Nuc inside, fits 4 frames. Photo credit: Ruth Rinehart

As it turned out, we got our first swarm on 04/23/2012. David managed to use a 17 foot ladder to get the Newport girls out of a tree. They were about 19 feet up. Two days later, Ruth captured the Wiley swarm.

Capturing a swarm is a very exciting event. People love to watch. The beekeeper gets to be the center of attention.

Before approaching a swarm without a veil, think carefully about whether Africanized bees are, or could be, in your area. Normally, honeybees are quite docile in the swarm state. But Africanized bees might be a whole different ballgame.

Illustration 76: Wiley swarm, 04/25/2012. Photo credit: Ruth Rinehart

Installation

Valerie was kind enough to invite us to join in on her package installation. This was four packages in a variety of equipment: Langstroth, TBH and Warré. A great educational opportunity.

Illustration 77: Package of honeybees. Photo credit: Valerie Gautreaux

Directions to Valerie's apiary bring us to a place a bit off the beaten path. That is not at all unusual. Beekeepers usually like their hives a bit remote and out of sight. It's one thing to provide bees for the environment, it's another for people to know who is providing them and from where they come. Not everyone is friendly about bees close by. I've seen pictures of hives that were shot as target practice. Not a happy event for the bees.

You can easily keep package bees happy by spraying them lightly with a little 1:1 sugar water every hour or so. Just spray it on the screen. Alternatively, brush it on with a barbeque type kitchen brush.

All newly installed bees are somewhat vulnerable in their new home. Entrance reducers are recommended to ensure that the bees can effectively guard their entrance before they have a chance to increase in numbers. Some folks also choose to feed the bees in this early stage as insurance against an inadequate nectar flow. If you are past the last frost, then feeding 1:1 sugar water would be acceptable and safe, however, you must not feed if you have supers installed for honey harvest (sugar water does not produce honey). Some also choose to feed pollen or pollen substitute. Feeding the sugar water is said to encourage quick comb building by stimulating the bees' wax glands. I have not chosen to feed my new installations. I prefer that bees find their food sources.

Installing Swarms

Swarms are the easiest when it comes to installing a colony of bees. They already have an accepted queen and they can be placed anywhere because there are no frames or comb involved. Having captured the swarm, installation begins with a colony of bees in a cardboard box or an official nucleus hive (without frames); the colony is sealed in a box of some sort. Swarms don't have a standard size, but you can expect a small swarm to be about 3000 bees, one pound, the size of a grapefruit, and a large swarm to be about 20,000 bees, the size of a basketball.

Open the receiving hive and remove a couple of frames in the center. That is where you will be dumping the bees. Open the container box where the bees are being kept. Dump the bees into the space in the hive like pouring them out of a bowl. You may have to shake the last of them out if they insist on hanging on. Or, if you have the time and are patient, you can just place the open container in front of the hive and let the stragglers find their way. You must ensure that the queen is in the hive and not one of the stragglers. If her highness is still in the box, then all the bees will be moving from the hive back into the box; not exactly what you wanted.

Illustration 78: Pouring a package into Langstroth. Photo credit: Valerie Gautreaux

Installing Packages

A package of bees is a ventilated wooden box which includes a can of sugar syrup for food. There are typically about three pounds of bees in the box, about 10,000 bees, and the whole thing weights 4-6 pounds shipping weight. There may or may not be a queen included, depending on what you ordered. If the queen is there, she will be in a separate little queen cage, with attendants. When you remove the feeding can, your next move should be to immediately lift out the queen cage. Tuck it in your breast pocket to keep it warm until you can give it your full attention.

A video that might be helpful follows. There are a number of things this beekeeper does that I would not do. Still, it's a good look at how package installation can be done. http://www.youtube.com/watch?v=YVW8CErDpjQ

One option is to pour the bees out of the package into the hive as with a swarm. A more gentle way is to use an empty deep below the target hive deep. Think Langstroth. The empty deep is on the bottom board. Place the package in the empty deep. Pull out the syrup can and grab the queen cage. Place another deep, with frames, on top of the empty deep. Install your queen between a couple of frames near the center of this deep.

End result from top to bottom:
Outer cover
Inner cover
Upper deep with frames & the queen
Lower deep empty other than the opened package
Bottom board

Illustration 79: Queen Cage. Photo credit: Valerie Gautreaux

Queen cages come in a variety of designs, however, they all have these characteristics: a queen with attendants, some way of installing the cage in a hive, and a way for the queen to exit the cage once she is established. Colonies not yet familiar with their new queen need four to six days to "get acquainted" before she can safely emerge. This includes "transit" time. Should she emerge before this adjustment, the workers will kill the queen. The queen cage might have a little metal piece sticking out. You can use this to hang the cage between frames like a tree ornament. Or, you can just press the cage into the wax on a frame. In any case, you need her to have good ventilation and screened access to the colony so they can get accustomed to her pheromone, keep her warm and feed her. Some queen cages have a candy plug, which the workers are supposed to eat, allowing the queen to emerge at about the right timing. Others simply have a door you have to pop open to allow her to emerge. Gently! That's your queen.

Now stick around and observe for a while. The bees in the package will move up to where the queen is located in the hive. No pouring or jolting necessary. Stay long enough to ensure that the queen has plenty of bees around her that can transmit her pheromone outward and attract the rest of the packaged bees. If, for some reason, the bees are not moving up to join the queen, then get the package back out and shake a few bees directly onto the frames with the queen. It is necessary that she have bees surrounding her for her survival. Be sure to put everything back the way you had it. That's it. Go home for a few days and let the girls all get used to each other. When you come back in a few days to let the queen out of her cage, all the bees will be out of the package. That's when you can remove the package and the empty deep along with the queen cage. All done.

Installing Nucs
A nucleus (nuc) hive is a small, three to six frames, fully functional hive, generally intended to facilitate transport. It's much more substantial, engineering wise, than a package, therefore, less likely

to be crushed. It has the advantage of including a few frames which allows the bees to actively raise brood even though this is a temporary home. Frames are typically standard Langstroth deeps. Some vendors ask that you return the frames or provide replacements. Others just include the cost of the frames in the product.

The cool thing about a nuc is that installation is a breeze. Simply remove the frames from the nuc and place them into your hive. Keep in mind that the brood nest should be roughly centered in the hive. Place the empty frames around the frames from the nuc. You can shake the remaining bees from the nuc into the hive or you can just set the nuc in front of the hive and let the girls find their way.

Obviously, this is great when transferring a nuc to a Langstroth hive because the equipment matches. If you are using TBH or Warré (or many others), you need to consider this before spending extra on a nuc. When nuc frames arrive with brood, bee bread and honey already in the frames, you will have difficult decisions about whether to modify those frames for your hive or to sacrifice their contents.

You MUST get the queen into the target hive. Wherever the queen is, the workers go.

When installing a nuc, this will almost never be a problem. Whether you see her or not, she will be on one of those frames you move into your hive. As an aside, queens in a nuc will usually be marked with a dot of paint on her thorax, so she will be exceptionally easy to spot.

You can test yourself on this, though, by simply waiting around to see where the worker bees are moving. If, for some bizarre reason, the queen is still in your nuc, then the workers will be moving OUT of the hive and back into the nuc. More than likely, you will be able to observe the workers moving from the nuc into the hive with purpose. That's a sure sign that there is a live queen in the hive.

My recommendation is that you don't spend your time trying to find the queen on the frame, but rather, spend your time observing the worker behavior which will tell you everything you need to know. Learning this will be valuable in the long run because eventually, you will get a queen that is not marked.

Disease, Parasite and Pest

© steve kennedy 2014

As of 6/12/2012, we thought the Quay St girls had come down with **chalkbrood**; pretty typical timing for this particular fungus, late spring or early summer. This makes me glad that my colonies are in diverse locations because that reduces the chances that the disease will spread. Chances are this fungus arrived via a drifting drone. I can reach this conclusion because I know the hive equipment was new prior to the Quay girls, so it would not have contained the fungal spores. Drones are known to drift, uninhibited, from colony to colony. It's part of natures way to ensure genetic diversity; spread the drones around geographically. Unfortunately, it has a down side when disease is spread by fungal spores.

This event is a nice segue into the domain of honeybee diseases and pests, a relatively advanced topic for beekeepers, but one that cannot be ignored forever. If you choose to be a beekeeper, you will, eventually, have to face each of the issues we are about to discuss. This is not an exhaustive text on honeybee disease and pests. It covers only some of the more common issues.

I practice and teach chemical-free beekeeping. That means the chosen response to a disease is generally mechanical in nature. In order to understand what you are up against, you need to learn about the causes and likely outcomes of each issue. This will give you insight about appropriate reactions, if any. For lack of a better approach, we will discuss these in alphabetical order.

1. American foulbrood
2. Chalkbrood
3. Nosema
4. Small hive beetle
5. Varroa mite
6. Wax moth

In all cases, good hygiene is a must to prevent spread. Whenever you encounter a pest or disease, be sure to clean your tools and clothing before manipulating another hive.

I choose natural, chemical-free beekeeping because I want to avoid the poison treadmill which will be discussed later in this text. As a permaculture style beekeeper, I actually take this a step further. I choose to not use even essential oils or natural acids. The idea is to allow the hive habitat to remain natural. My hope is to find the bees that have the genes that will be successful in my habitat and then propagate those genes. This is a work in progress, not a done deal yet.

American Foulbrood

Warning! Extremely hazardous! **American foulbrood (AFB)** is the most serious bacterial disease of honeybee brood. This is the killer you really don't want. It's deadly, spreads easily and remains indefinitely on your equipment and in the environment. You can't see it, so you can't avoid it, which really stinks, figuratively and literally. Young honeybee larvae ingest *Paenibacillus* larvae, a rod-shaped bacterium, which is visible only under a high power microscope. Spores germinate in the gut of the honeybee larva. Infected honeybee larvae normally die after their cell is

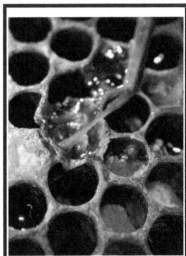

Illustration 80: American foulbrood rope test. Photo credit: public domain, USDA

sealed. Each dead larva may contain as many as 100 million spores.

One important symptom is cappings that are perforated and sunken into the cell. When you see this, you need to do a test for ropiness as seen in the photo on the previous page. Using a toothpick or equivalent, poke the cell, spin the stick and remove. If the sticky cell content ropes out like a string, then this confirms AFB. Colonies with AFB will have a foul barnyard odor.

After AFB damage is done, you can see the resulting "scales" and this can be very helpful in confirming your diagnosis when you are making a decision about destroying the equipment. You can see a great picture of an AFB scale here:

http://beeinformed.org/2012/08/detecting-american-foulbrood-with-a-blacklight/afb-scale/

and see a picture of confirmation of AFB using a black light here:

http://beeinformed.org/2012/08/detecting-american-foulbrood-with-a-blacklight/afb-under-blacklight/

AFB is best prevented rather than overcome. Chemical free prevention involves beekeeper hygiene and hygienic behavior in the bees. Terramycin is used by those who treat, but as is usual with drugs, bacterial resistance is underway.

You can innocently introduce AFB by feeding your bees store bought honey. The spores are transmitted in honey in viable form, though not a risk to humans. Never feed your bees store bought honey. If you must feed using honey, always use your own so you know it is disease free.

Because you really don't want to get this, you also really don't want to give it to others. When diagnosed, sadly, the infected equipment should be destroyed by burning. That's everything, the entire hive, including any remaining live bees, and preferably anything that has come in contact with the disease, like gloves and clothing. Get a large, hot fire started in a pit where you will bury the remains. Approach the hive after dark (all bees are home), open and mist with soapy water (prevents flying). Disassemble and stack on the fire. Do not try to save anything, including honey.

Chalkbrood

Chalkbrood is an infectious fungal disease which is caused by *Ascospaera apis* infesting the gut of the larva. It rarely kills the colony, but impacts the worker population, thus weakening the colony and reducing honey production. Spores are highly infectious and carried in

Illustration 81: Chalkbrood Mummies. Photo credit: Jeff Pettis, from wikipedia

contaminated pollen by foraging bees, drifting bees and drones. Exchanging equipment and bees, feeding contaminated honey and using contaminated tools and gloves are all potential activities that can spread the spores.

Larva ingest the spores when fed infected pollen. *Ascospaera* grows best when the brood is chilled or the colony experiences an increase in carbon dioxide (CO_2) levels. If nurse bee numbers become insufficient, the brood may be left unattended giving the spores the opportunity to grow. The first larvae to be affected are those around the edges of the brood. After germination, the vegetative growths (hyphae) of the fungus invade the larval tissues of capped brood and kill them. Dead larvae become chalky white and fluffy and swell to fit the cell, then shrink, harden and become mummies. Workers remove the mummies from the hive and they can often be seen on the hive floor and outside the hive.

Because of the temperature and ventilation aspect of the disease it is more likely to occur in small colonies or nuclei.

One viable response is do nothing; let the colony overcome the infection on their own or perish. To help with temperature control, consider increasing the nurse bee population by introducing a frame of brood covered with bees from another colony. Another option is to increase the ventilation to help the CO_2 level. A more invasive option is to pitch in and help with some house cleaning to reduce the fungal load; sweep out the mummies and bump uncapped comb loaded with mummies to knock the mummies out. Heavily affected comb should be destroyed. Another natural treatment is acetic acid, however, I do not consider that to be chemical-free beekeeping.

Nosema

There are two varieties of nosema: *Nosema apis* and *Nosema ceranae.*

Both species are a unicellular fungus in the gut of the bee. Spores are passed within the bee's waste. The bees have a tummy ache.

Illustration 82: Evidence of nosema. Fecal spots small and large around entrance. Mild case. Photo credit: Don Studinski

Nosema apis is the one we notice. It usually shows up in winter or early spring when the bees have been cooped up for a long time. The beekeeper starts seeing brown spotting on the outside of the hive, especially on the front porch. The poor things just can't hold it any longer. Spring build up is delayed. Usually, the colony survives and the proportion of infected bees begins to decline rapidly. This decline occurs because the feces are normally voided away from the hive when regular flights become possible as the weather warms. Since the old bees now die off and are replaced by healthy bees emerging from the brood combs, the disease may not be detectable in the colony by the end of the season.

Nosema ceranae is nasty because it shows no symptoms and can cause rapid decline and collapse with CCD like symptoms. This one can get the bees at any time of year and is tightly coupled to neonicotinoid sub-lethal exposure.[24]

---

24 http://www.ncbi.nlm.nih.gov/pmc/articles/PMC2847190/, visited 4/1/2014

Those that treat use Fumagilin-B, an antibiotic. I choose not to. Consider swapping out the bottom board so you can wash off the feces. Feed them fondant to enure they are not doubly stressed. Hope for the best.

Small Hive Beetle

Rarely seen in Colorado, newly emerged adult small hive beetles are light brown, becoming progressively darker (almost black) as their exoskeleton sclerotizes (hardens). Naturally occurring small hive beetles can vary greatly in size, possibly depending on diet, climate, and other environmental factors.

Small hive beetles seek out cracks and crevices where they hide from bee aggression. Guard bees are posted to keep the beetles from accessing the brood comb. Amazingly, the beetles are able to stimulate their guards to feed them using beetle antennae to rub the guard bee mandibles.

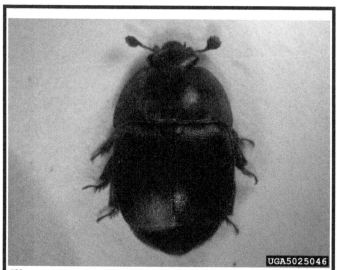

The beetles seek the brood comb because it contains all the food stuffs needed for beetle reproduction. A female small hive beetle may lay 1,000 eggs in her lifetime. The eggs are humidity sensitive, dying in air below 50% humidity which may explain why they are so rare in Colorado. Larvae feed on honey, bee bread and brood causing significant damage including fermentation. Honeybees may abscond when adult beetle population exceeds 1,000 in the hive.

Illustration 83: Small hive beetle. Think 1/4 inch long. Photo credit: James D. Ellis from wikipedia

Chemicals are available to control the beetles in the hive and in the soil where they pupate. Obviously, I would not recommend it. I have seen this pest in Colorado due to packages and nucs purchased from out-of-state. Like anything in nature, should the population become numerous, the species that feeds upon that population will soon arrive to bring the population back into control. Should I have a need, I would be finding a way to quickly extract any harvested supers and seek storage areas where humidity can be controlled.

Varroa Mite

Varroa destructor, the varroa mite, can only replicate in a honeybee colony. This parasite attaches to the body of the bee, sucking **hemolymph** (blood), leaving open wounds which are vulnerable to infection, weakening the bee and transmitting viruses in the process. Infestation can cause colony death, generally late fall to early spring. The adult mite has eight legs on a relatively flat reddish brown body.

Illustration 84: Varroa mite. Photo credit: public domain, USDA, from wikipedia

These mites reproduce on a ten-day cycle. The eggs are laid on honeybee larvae (preferring drones). Within the cell where the larva is growing, several females and one male hatch. If each cycle creates four mites for every one adult, then 100 mites become 6400 in one month. In late fall, when honeybees quit raising drones, the mites must use worker larvae which can lead to a precipitous loss of workers and may lead to collapse.

Of course, our early response was to treat with chemicals which many continue to this day. Also of no surprise, the mites that survived are chemical resistant and now they have multiplied. We switched chemicals. Now the cycle is repeating. But at least the big chemical companies are doing well (yes, that was sarcasm).

Screened bottom boards allow mites that fall to drop all the way to the ground where they have no chance to regain entrance to the hive. Some say sprinkling pulverized granular sugar (not powdered sugar) causes the mites to lose footing and drop off the bee.

Small cell brood comb is a huge topic unto itself and will not be covered here, however, the topic is related to varroa in that some beekeepers believe that by allowing the bees to return to their natural, smaller, cell size the bees gain ground against the mite. The mite preference for larger drone cells suggests that this notion may have merit. This idea is also appealing from a permaculture perspective.

Drone frames have foundation specifically sized for the bigger drone cells preferred by varroa. When used in the brood area, this encourages the queen to lay plenty of drones which will attract the mites. The idea is that an alert beekeeper will regularly remove the drone frame and pop it in the freezer to kill the larva and mites. Replace the drone frame to allow the housekeepers to harvest the dead and start the process again. The interruption in the mite brood cycle can be very effective at reducing mite population. The TBH equivalent is to regularly harvest (cut out, cull) the drone brood positioned on the outer edge of each top bar's comb.

Another approach to brood cycle interruption is using the split process after the summer solstice. By forcing the colony to raise a new queen, thus interrupting the honeybee brood cycle, the beekeeper forces the mites to wait an extended period for access to honeybee brood where mite eggs can be laid. When honeybee larvae finally reappear, the adult mites overwhelm the few available honeybee larvae

with too many mite eggs. The mites starve for lack of food and a whole generation of mites is killed. The mites don't have enough time to rebuild their population before winter, but the honeybees continue to expand and have a healthy colony going into winter.

Finding and multiplying honeybees with hygienic behavior that successfully defend the colony against their natural predators is, in this author's opinion, the most likely strategy for long-term success.

Wax Moth

Very common in most parts of the world except the colder regions, the Lesser wax moth is a small moth, the Greater wax

Illustration 85: Ruined comb held together by wax moth silk. Photo credit: Don Studinski

moth is somewhat larger. Our concern is not so much the moth, but their larva, also known as waxworms, which are widely bred as food for pets. The larva eat beeswax, especially brood wax, seeking honeybee larvae, shell casings, bee bread and honey as they tunnel through the comb leaving a trail of icky silk and waste. This is nature's way of recycling that wax and we would normally be grateful for the service except when they have invaded a living colony. In most cases the worker bees will eliminate them and keep the moths from over-running the colony, however, when weakened, a colony can lose this battle. Beekeepers can help a colony keep the pest out by removing the debris that accumulates on the bottom board or in cracks and crevices. Braconid wasp is frequently found with wax moth. It is a predator of the moth ... yes, those wasps can be useful!

When faced with a scenario where you need to store brood comb, these troublesome worms can be controlled by freezing the comb, killing all eggs and larva, just prior to storage and then storing the comb in sealed containers. Leave the comb in the freezer long enough to ensure thorough freezing, like overnight. You can also use the other end of the temperature spectrum; temperature of 115°F (46°C) for 80 minutes or a temperature of 120°F (49°C) for 40 minutes will kill lesser wax moth. However, be careful, temperatures exceeding 120°F may sag wax, defeating your purpose.

Illustration 86: Greater wax moth. Photo credit: dhobern, from wikipedia

Beekeepers that use chemicals may choose to use paradichlorobenzene (PDB) on stored hive boxes. However, notice the warnings on the label about possible injury to humans and other life forms. Did I mention that I do not use chemicals?

May ... Building Comb

Most of our swarms for the year will come during May. Swarms ramp up during April and ramp down during June. There will be a few widely scattered swarms in July. The main action happens in May. After having chosen a new home, the first order of business for a swarm is to build comb. We prefer that they choose to live in one of our swarm traps or one of our hives. But, no matter where they choose to live, they must have comb before they can store any food and before they can raise any new brood. Comb is a big deal. It takes a lot of energy and time to build. Beekeepers must treat it as the precious asset it is.

Illustration 87: Boulder swarm, 5/6/2012. Photo credit: unknown, public domain

If you have a strong colony and you have not yet put on your supers for honey harvest, then stop reading now and go do it. Even if you don't get honey, you can at least get the bees to help you by building comb on those super frames.

Brood Pattern, Managing Comb

Today is May 1, 2012, and most of us have honeybees installed in a hive by now. Some have over-wintered successfully, others are new from a package, a nuc, a split or a swarm. In any case, the next thing we need to do is verify that the queen is laying a good brood pattern. This is an inspection for which we must not delay. Verification of a healthy, laying queen is essential. Her failure could spell failure for the whole colony. Before she can lay, she must have comb. Some thoughts on comb management are in order.

A new colony's first task is to build comb. No comb means no place to lay eggs and no place to store food. In some cases, you will have provided comb from previous seasons. This saves the bees a lot of effort. Historically, beekeepers have made every effort to save the comb for their honeybees to use for many years. These days, the wise beekeeper will accept that there are poisons building up in those combs every year. They need to be rotated out on a regular basis. I'm rotating comb out after three or four years of use. The 2012 Bee Informed survey[25] indicated that those that rotate out old comb are reporting just as many colony failures as those that do not. Beekeepers need to continue to monitor these results and keep an open mind about modifying

Illustration 88: Great progress on newly built comb on Warré frame. Swarm 4/25. This photo 5/17. Photo credit: Ruth Rinehart

25 http://beeinformed.org/2013/10/brood-comb-management-and-treatment-of-dead-outs-national-management-survey-2011-2012/

practices as the preferred method comes to light.

Illustration 89: Boulder, CO swarm, 5/6/2012. Photo credit: unknown, public domain

Rotating comb out is straight forward in the case of Warré hives. We introduce a new deep at the bottom of the stack. The ladies will use this deep for brood first. Brood comb will be relatively thick because the wax is coated with propolis, giving the comb some substantial structure. It doesn't break as easily as honey comb. As the months and years pass, this deep will migrate slowly up the stack; new deeps being added to the bottom as necessary. Eventually, the deep reaches a height that the bees transition the comb from brood to honey. At that point, we might start referring to the same box as a super rather than a deep. The bees want the honey over their head, that is, over the brood nest or cluster. The heat generated by the cluster in winter will rise and soften the honey allowing consumption. If the colony generates enough honey, then the highest super is the beekeeper's harvest. That's the point at which the comb is rotated out. We cut the comb off the frame for harvest. In this manner, comb may be rotated out earlier than the three to four year target, but will almost certainly not be pulled later. Bees that have not produced harvest in three years would probably be replaced with bees that do. The age of the comb should be taken into account upon such an event.

Illustration 90: Capped brood had to be laid at least 9 days ago. Photo credit: David Albrecht

Proud first year beekeeper, David, shared this photo with our group. This is an example of some newly drawn comb. His colony was delivered as a package in northern Colorado just 11 days earlier. What we can infer from this picture:

1. The delivered queen was, in fact, ready to lay, as advertised.
2. The girls were able to draw comb enough to use for eggs in two days or less. Brood gets capped on the ninth day.

Langstroth hives could be done exactly like Warré if you choose. More typically, Langstroth beekeepers add deeps and supers to the top rather than to the bottom of the stack. My style is to use three Langstroth deeps, with medium supers above. In this case there are two sets of boxes to consider for rotating wax out, the deeps and the supers.

Illustration 91: Diamond Court swarm. Photo credit: Robin Guilford

Each year, you want to take advantage of a time when one of the deeps is essentially empty so you can eliminate that wax without killing brood or wasting honey. The most likely time for this is at first inspection. The colony will have moved up during winter consuming honey as they go. They will be found in the top deep or the top two deeps. This will leave the bottom deep unused with no brood and no honey inside. That's the time to pull the frames and replace with fresh. Move the middle deep to the bottom board, the top deep becomes the middle and the new frames become the top deep.

Where the supers are concerned, wax rotation is not an issue. No brood is raised in that comb and it's only on for a short time each year. I would reuse that comb as much as possible.

Top-bar hives require a different management approach and, therefore, do not fit the pattern for Warré or Langstroth. In this case, harvest is likely in spring. To ensure the colony has adequate stores to survive winter, leave the honey in place over winter and pull the frames for harvest in spring. Rotating out comb is made straight forward by sliding top bars as through a queue. Harvest the honey farthest from the brood nest, then slide all the remaining bars toward the space where the harvested bars were removed. This will have shifted the brood nest. Place the empty harvested bars at the opposite end of the queue. The bees will expand the brood nest in that direction building new comb as they go. Comb is retired as a result of

Illustration 92: Sable swarm. Photo credit: Don Studinski

honey harvest. The comb will likely never get more than three years old for the same reasons as stated for Warré.

Whether building new comb or reusing existing frames, the bees need to get moving on raising a new brood which will replace the aging workforce. What we want is a vigorous queen laying a lot of eggs in at least a football pattern if not a full frame at a time. Her laying should uniformly fill the space she is using. Nearly every cell should be full. Depending on the timing of your inspection, you may be looking at eggs, larvae or capped brood. In any case, you want to see all the cells filled in the brood area. A spotty pattern could indicate a problem. A failing queen may be due to age, injury or inadequate mating. Whatever the reason, if you don't find the pattern you seek, you should initiate queen replacement.

Illustration 93: Sable at a distance. Photo credit: Don Studinski

Give the queen two weeks after emerging to start laying. The first week is for her to mature sexually. She emerged as an adolescent. She will fly as an adult. The second week is for sex. She will mate with 10 to 20 drones in drone congregation areas (DCAs). A drone's sex organ is torn out upon mating. If this gets stuck in the queen, then she must return to the hive to get help from the workers to remove the organ. If it falls away spontaneously, then she may mate with more than one drone in the same flight. If you are eyesight challenged, like myself, you would benefit yourself and the bees by waiting a third week to peek for a good brood pattern. In that case, the eggs will have hatched and the larvae grown large enough to see easily in the bottom of each cell. I prefer to disturb the bees as little as possible, so I wait the extra week. But, if you can see eggs, you can check at the beginning of the third week. Of course, if you don't see the eggs, then you may have to check again. You don't want to replace her unless you must. This has to be balanced with the motive to replace her as quickly as possible.

Illustration 94: Sable up close. Don't transport in sealed container. Photo credit: Don Studinski

Whether your new bees are a swarm, a package or a nuc, it is time, or near time, for you to get out there and inspect for a good laying pattern. This is verification of a healthy queen. Do not confuse seeing the queen with verifying her health. You must see the laying pattern with reasonably solid brood, to verify her health.

In the past, I have successfully moved smaller swarms in a Coleman ice chest with the lid on. I've made it with a swarm in transit for about an hour and they were okay. With the poor Sable Street

Illustration 95: Sable new comb. Clearly, they lived. Photo credit: David Braden

swarm, we were driving for about half an hour and they were overheated. This swarm may make it, or may not. I would recommend not using an air tight box like this. This is why you want a nuc for transporting swarms. You need ventilation even for short periods of time.

Concerning Drones and Mating

This is mating season. All the colonies are producing drones as their contribution to the gene pool. This year is proving to be particularly happy for the honeybees. They are producing a large number of

Illustration 96: Oak Swarm. Photo credit: David Braden

Illustration 97: Oak Installation. Photo credit: David Braden

swarms, which implies a large number of virgin queens. Some virgins inherit the hive. Mom has flown with the swarm. Some virgins fly in a **secondary swarm**, a swarm with a virgin queen. All virgins need to mate. Should a colony throw more than one swarm, the second and following swarms include a virgin queen which must fly to mate. These are also called **after swarms** or **virgin swarms**. Every mating event results in the death of the drone, just as a sting event results in the death of a worker. Each virgin will need 10 to 20 drones as mates. Therefore, the bees need a large number of drones to provide this service. Mating actually happens in the air, not in the hive. The drones from many hives create what is called a drone congregation area (DCA) 40 to 60 feet in the air. This is where mating occurs. The virgins fly to the DCA where they are approached by the most vigorous drones who chase the queen for the opportunity to pass their genes to future generations. Because queens will mate with 10-20 drones, technically, they may not be a virgin at the time of mating, but we do not generally talk about that subtlety.

Because the drone's phallus is ripped out of his body after mating, it may get lodged in the queen's body. Generally, it will fall away while she is still in flight. Sometimes, this doesn't happen and that requires her to make another trip back to the colony to get the workers to remove the debris for her. Then she can fly again.

Interestingly, drones are not considered a threat to a colony, even when they are not a brother. This allows drones to "drift" from hive to hive which has a serious down side because it facilitates the spread of parasites and disease. As humans, we think the bees should prevent this in order to improve the overall health of the population, however, nature must have a reason for allowing this behavior. I suspect this is one way nature culls the weak. Or,

Illustration 98: Lafayette, CO swarm, 5/2/2012. Photo credit: unknown, public domain

perhaps the ability to spread genes around geographically is more important than not spreading viruses and bacteria. In any case, we need them all, even parasites, viruses and bacteria. Each has a gift to give to a healthy ecosystem, even if we don't understand what it is. Millions of years tells us that this strategy is successful.

Identifying Laying Worker or Drone Layer

It's not necessarily a problem to be seeing many drones moving in and out the front door, especially this time of year. However, it is good to be alert about what you observe and seeing many drones is a good prompter for you to go ahead and perform an inspection to ensure you don't have a dead queen and laying workers which can only produce drones. A laying worker is sometimes referred to as a **drone layer**, but that term can also apply to a queen that was not adequately fertilized. You want to see a good brood pattern and capped workers. If all you find in there is capped drone, in a shotgun pattern, then you need to requeen.

Illustration 99: Laying worker evidence. Photo credit: Susan Sommers

The queen pheromone inhibits workers from being stimulated to lay eggs. If she is dead or missing, the pheromone level drops and will, eventually, allow workers to start laying. Workers, because they cannot mate, are only capable of laying eggs for drones (haploid). This is a very "ill" colony. Laying workers must be eliminated to get the colony back on track with a healthy queen.

Once you have a laying worker, the colony thinks they <u>have</u> a queen. Putting a new queen in will just be a waste of money and a waste of the queen because they will kill her. Not every colony can be saved and once things get to the laying worker stage it is a dead end for that colony. You can do one of two things, either shake the queenless colony out in the weeds or combine them with your queen right colony. In the shake case, shake all of the bees out of the box and onto the ground away from the original hive position. About 50' should suffice. When you shake them out there are two possible follow-up procedures. The first is to completely remove their hive forcing them to either

Illustration 100: Drone layer results. Photo credit: Susan Sommers

beg their way into another colony or perish. Some say that whatever colony the laying worker returns to will kill her. The second is to return the now empty hive back to its original location and the field bees will return, but not the laying worker. Then re-queen the hive.

We will soon see examples of healthy brood patterns on Warré, Langstroth and TBH frames.

Here we have an example of a colony in trouble. These frames show the widely scattered, bullet shaped drone cells typically seen as a result of a colony having gone queenless. This is what you would see from a laying worker. This particular case is unique because it was not a laying worker. This was a package queen that had not been successfully fertilized. We found her and destroyed her.

Having intentionally killed the queen, all the bees in the colony know they are queenless within four hours. Queenless bees will normally attempt a supersedure to replace her; however, in a case like this, they have no larva that could be a queen. Therefore, we must requeen using a caged queen. We introduced a new queen in a cage, allowed the colony to get used to her pheromone for four days and then released her.

Some caged queens come with attendants, also called a **retinue**. The colony may or may not kill the attendants. All the bees become one family with the new queen pheromone over time.

Keep in mind that <u>a queen is a colony and a colony is a queen</u>. By remembering this saying, you will be reminded that the loss of a

Illustration 101: Forest Street swarm. Photo credit: John Brown

queen is the loss of a colony. Her genes can no longer be added to future generations. When her living daughters and sons die, that's the end of that line. What this means is that when requeening this colony, we did not save the colony. The last of their survivors helped a new colony get started with their new queen. That's the best we could do. We lost a colony and we gained a colony in the same hive.

Queen Difficulties

A drone laying queen needed to be replaced. Our first attempt to diagnose the issue included a search for the failing queen, but we could not find her and made the (wrong) assumption that she was not there. We figured a laying worker was the source of the drone cells we were seeing. After having purchased and seasoned a new queen for four days, we released her into the hive. But, something didn't seem quite right at the time. We waited one week and checked for brood ... nothing. Oops.

Upon closer inspection, we found the queen. Not the queen we just released a week earlier. Nope, this was the original. The new queen, of course, was immediately killed by the workers loyal to the existing (and failing) queen. That's why something seemed amiss when we released the new queen.

Illustration 102: Louisville, CO tiny swarm. Photo credit: Don Studinski

After killing the failing queen, we left the colony definitively queenless ... intentionally. We wanted them to have at least 4 hours so they would all know they were queenless. The next day, we got another new queen and installed her to begin a four day seasoning. This time, the whole process went as expected. An expensive lesson.

Why did they not replace the original failing queen?
Answer:
She never gave them a chance. She has never laid anything but drones. There is no viable egg with which to make a new queen. Still, they did build loyalty toward her despite her failure.

Will they be able to increase enough to make it through winter?
Answer:

Yes, once we get a healthy queen in there, they will be fine. It is still early in the year.

Are there laying workers?
Answer:
There are always laying workers. As long as there is a queen in the hive, the laying workers are kept under control by normal queen and brood pheromone. Queen and brood pheromone inhibits workers from laying. A hive has to be queenless for several weeks for workers to start laying successfully and for nurses to bother raising the drones. Actually, that's what is so interesting about this case. It has all the symptoms of a laying worker, however, the failing queen was there the whole time.

Queen Health Inspection Timing

It's important to verify queen health. There are a number of scenarios that can happen and each one calls for, potentially, different timing. Beekeepers need to be familiar with the logic that goes into making the decision about when to inspect.

Illustration 103: Braden queen, Italian, from a nuc. Photo credit: Don Studinski

Consider over-wintered colonies: Superior and Birch. When we did first inspection on Superior we found them with a tummy ache ... this is known as nosema. We knew this because of the poop marks at the front door. We did nothing to treat them other than replace the bottom board. This gave them a break from any accumulation of feces on that bottom board, and also gave us a way to monitor their health moving forward ... watching for the absence or presence of more poop marks at the front door. Today, they are a super healthy colony with dozens of bees moving in and out every second. They are producing honey. The point to note is that we *did not* look back inside that colony at all. We monitor health from outside and let them manage inside. In the case of Birch, once the three-way split was completed, we also have not looked back inside that colony. Again, they are super healthy with dozens of bees flying in and out every second. In this case, leave them alone. Both these colonies did get a third deep, but that did not require frame by frame inspection.

Next, let's consider the case of a newly purchased nuc. The queen should be already laying upon delivery. These bees are the daughters of this queen. One could reasonably observe the laying pattern upon installation and leave them alone after that. One could also reasonably come back and have a look a couple weeks later just to ensure that the move didn't upset or injure the queen. After that it's time to leave them alone. Monitor health from outside. Let nature run its course. This case could have been applied to either the Braden or the Rinehart colonies.

How about a package? In this case, the queen is fertile (supposedly), but not related (genes) to the workers in the colony. We generally give her a few extra days in the cage to ensure "acceptance" by the colony. Once released, assuming she has comb to work, she should start laying almost immediately. A couple of days to settle in may be required, but nothing more. How soon can you verify her laying? Here are the factors that affect the answer. Does she already have comb in which to lay? Can you see eggs or newly hatched larvae? If the colony has to draw comb, then they will need at least 2 days before they will have enough comb in which to lay. Once she is laying, you will be able to see capped brood on the 9th day. So, the factors you must know include when do you think she will begin laying and how good is your eyesight. You will want to adjust your inspection timing accordingly and it's probably wise to error on the side of a longer wait rather than too short. This is because if you cut it short and don't see what you want, you will need to come back and disturb the bees with a second inspection. If you wait longer and you don't see what you want, then you know you have a problem that needs your attention.

Illustration 104: Braden queen lays in a nice football pattern, center cells already emerged. Photo credit: Don Studinski

Then there's the case of swarms. Swarms come in two flavors, "**primary swarm**" and "secondary swarm" (or virgin swarm). The first swarm thrown by a mother colony, the primary swarm, includes the old queen. This is the second (and potentially) last time in her life that she will fly. The first was her nuptial flights. She is fertile and healthy or the colony would not have swarmed. Given two days to draw comb and begin laying eggs, and another nine days for larvae to be capped, you can expect to peek inside and see capped brood in 11 days. Since my available time comes in week increments, I always plan to peek after 14 days ... if I'm sure she was not a virgin, that is, the swarm was large, 5 - 7 pounds, the size of a basketball or bigger. In the case of a secondary swarm, the queen is a virgin. The mother colony already threw the old queen out and they were still healthy enough to multiply. They made an additional queen and proceeded with a second or even a third swarm. The important point here is that a virgin needs another two weeks before she will start laying. The first week is for her to mature sexually. The second week is for her to fly and mate ... a dangerous venture in and of itself. Swarms football sized or smaller and swarms later in the year are more likely to contain virgins and the beekeeper must adjust the timing of inspection accordingly. When I think it may be a virgin swarm, I time my inspection about three weeks after installing them into the hive. At that time, I should be able to see at least large larvae and, more than likely, capped brood.

Illustration 105: Newport queen, a mutt. Photo credit: Don Studinski

If you are inspecting and you don't know what you are looking for, then why are you looking? Don't disturb your bees more than necessary. However, necessary may include "I just want to look for my own education." For example, suppose you want to know if it was a primary or secondary swarm. A good way to verify this is to inspect after just one week. In that time, if you are already seeing larvae, then the queen is obviously already fertile and it must have been a primary swarm.

When you inspect bees, you will kill some. You will eventually get comfortable with this. If you choose to inspect without smoke, you will kill more than if you use the smoker. Angry bees sting and those that sting die, even if they leave the stinger in your clothing and not in your skin. Once your colony is 50,000 or 60,000, there is no way to work with them without killing some. I like the phrase "you can't save them all" even while I'm trying to save as many as I can. For example, Friday, I was catching a

Illustration 106: Newport brood. Not as solid as desired. Photo credit: Don Studinski

swarm in the cold. Bees were on the ground. I waited patiently for about an hour for them to find and climb into the box. Perhaps a silly use of my time over maybe 50 bees. Still, I will do that from time to time. On the flip side, when I place a super on a hive and squish some workers that are in the way, I don't worry about it. You can't save them all.

Queen Health Inspections

Having installed a nuc, package or swarm, your adventure as a beekeeper has begun. Soon thereafter, you need to confirm that your queen is performing properly. This is nature and there are no guarantees. If you don't check up on her in a timely manner, you may well be disappointed when you do get to it. The laying worker problems are more difficult to correct than the failing queen found early. The following series of photos shows queens and brood frames that look good. Remember, seeing your queen is nice, but not

Illustration 107: Rinehart larvae. Photo credit: Ruth Rinehart

necessary most of the time. Seeing your brood is totally necessary.

The Braden girls came from a purchased nuc. The queen is Italian. Her performance in May of 2012 is pictured on an earlier page. The football shaped brood pattern is good. The middle empty cells is where new workers have already emerged. This is what we seek when we want to verify a healthy queen: good brood pattern where laying starts in the middle and spirals out from there.

The Newport queen was performing quite well as of May, 2012.

The queen is nearly center in the photo. She has some large golden bands on her abdomen near her wings and about three darker bands that stick out beyond her wings. She is oriented abdomen to the left, head to the right. Also of note in this photo is larvae in the cells under the queen and around that area. Look for small white glistening worms.

In the Newport brood photo there are several things worth noting. Upper left you will see capped honey. Below those pretty white caps we find cells filled with nectar that has not fully cured, that is, it's not honey yet. We call this uncapped honey.

The pretty colors in that rainbow shape is pollen mixed with enzymes and honey, also known as bee bread. Bee bread will show up in a wide variety of colors, which is a good sign that a diversity of forage is available to the bees. That's what they need to be healthy.

Below the bee bread layer is our brood nest. Some of this brood is capped. That's the tan colored caps. In this case, all the capped brood is workers. Capped drone and capped queen cells look significantly different. As you can see, brood comb becomes discolored almost immediately. Compare it to the newly built, white, comb that is on the outer edges. When comb is new, it's a beautiful white, like snow.

Illustration 108: Rinehart queen's wall to wall capped brood. Photo credit: Ruth Rinehart

We also have a nice photo of queen Rinehart's larvae. Look in the bottom of each cell. That little white worm is a honeybee larva. If you can see one that is *smaller* than 1/6th of the bottom of the cell, then that is a candidate to become a queen. An egg gets laid and it hatches into a larva on the third day. All female larvae get fed royal jelly an average of 1300 times each day on days 4, 5 and 6. During days 4, 5 and 6, all female larvae are candidates to become a queen, but on the 7th day, the decision is finalized. If the larva is to become a worker, then the diet changes to bee bread and honey. If the larva is to become a queen, the royal jelly feedings continue for days 7, 8 and 9. Those additional royal jelly feedings are what makes the difference in a fully developed reproductive system. All female larvae are capped on day nine.

Queen Rinehart is not satisfied with a football pattern, thank you very much, she will use the whole frame. Awesome.

Queen Victoria at Linda's house, from a California package, has what we will call a "creative" family. Despite our efforts to present them with parallel frames on which to build comb, and despite the fact that we rubbed beeswax in a line on those frames suggesting where to build, they clearly know better how to build in their space than do we.

Here is what to notice:

We pulled a far edge frame which appeared to have the least comb attached. We were trying to free up some room with which to work. What we found was several combs attached all running diagonally, therefore, they all broke as we lifted the frame. The tiny comb to the right came out whole, but fell off shortly after we lifted the frame. You can see the remnants of four other combs as you look left across the frame.

This is a worst case scenario because there is *no way* to lift comb out of the box for inspection without breaking it. If we force our way into this comb to see the brood, we will destroy much of the built comb and brood, and we will run the risk of killing the queen in the process.

Lifting the box over head, we could see the comb running diagonally through the box. It's beautiful comb, built just the way the ladies want it, but very inconvenient for the beekeeper.

Here's why I find these girls to be "creative." Diagonal isn't enough to satisfy them. They also had to show how to make "S" curves half way through the box.

We were able to see some capped brood near the bottom edge of one frame. We decided that would have to do when it comes to verifying Queen Victoria's health. Our reasoning is that if we see capped brood near the bottom, then it's likely there is more capped brood running up the frame.

Illustration 109: Victoria girls force us to break comb to see a frame. Photo credit: Linda Chumbley

Deeps and Supers

This year, we performed first inspections and identified who needed to be split. We executed our split process. We received our nucs and packages and got them installed. We received nature's abundance in the form of swarms and got them installed. We performed queen health verification after a carefully calculated wait unique to every colony so we could see

Illustration 110: Victoria diagonal comb. Bummer. Photo credit: Linda Chumbley

capped brood in the correct pattern. So far, it has been an excellent year.

Our next focus is getting the correct number of deeps and supers on each colony such that they have an appropriate amount of space for brood and honey. Appropriate brood space for a small swarm (grapefruit sized or smaller) might be a single deep or even just a nuc box. They need enough room to expand their population; however, if we over-do it, then there might be too much space for them to keep warm and / or too much space for them to defend. Appropriate brood space for a vigorous colony that is harvesting honey is two or three deeps. The Birch Street girls are filling three deeps with brood now. They also have three supers on for honey harvest. They are on track to be at about

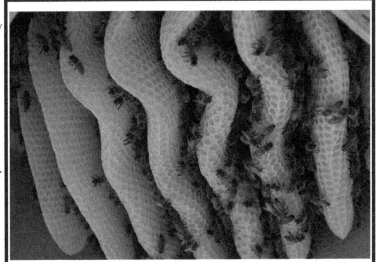

Illustration 111: Victoria "S" shaped diagonal comb. Photo credit: Linda Chumbley

100,000 bees by July. I make it my mission to get every colony to three deeps. I want a strong healthy bee population going into winter, and on the other end, I want them to have plenty of room for expansion come spring. It's May of 2012, and I'm already thinking about preventing swarms in 2013. Appropriate supers for honey will depend on the worker population for each particular colony. Queens that produce a large workforce give the colony a better chance of producing honey for harvest. Therefore, my first focus is to give her all the brood room she will use, and only after she has shown that she will use all three deeps will I give them honey supers. Does this mean that every frame of every deep must be full of brood? No. I'm watching the front door for traffic. If you have provided three deeps and you are seeing about a dozen bees per second arriving and leaving the front door, then you are ready for supers. I will sometimes put on supers earlier; however, be careful that you don't set your colony up to provide you with harvest in supers and leave themselves short of honey in the deeps. For example, on a strong colony that I know has time to ramp up to 100,000 bees, I'll put on supers in April and try to catch some of that early flow. However, on a small swarm that will not likely show their true colors until the following year, I will probably not put on a super at all in 2012.

Preferring not to have to scramble to get equipment at the time that it is needed, I choose to purchase new (to me) equipment in advance. This year, I have purchased several hives second-hand from beekeepers that I know perform chemical-free beekeeping like myself. I don't want the chemical residue that may be present otherwise. I have also purchased some equipment new. It's worth the effort of talking to your fellow beekeepers to see who might be selling what whenever you are in a purchasing situation. For example, I have seen a new Langstroth 10-frame deep with frames & foundation for sale at $40. I've also seen that same setup for sell at $75. Both right here in Denver. Don't expect to find that low price every time; just be aware that it pays to shop.

Illustration 112: Dudley Street swarm. Photo credit: Don Studinski

Illustration 113: Thornton, CO huge swarm out of reach. Photo credit: Don Studinski

You should be planning now for what, if any, additional equipment you are likely to need in 2012 for the colonies you already have. In June, we will start planning for additional equipment for summer splits we may undertake.

Adding to the bottom of the stack is a Warré management method with which I am experimenting. I'm generally using that technique with my Warré hives and with my Langstroth hives. However, I'm not locked into it. For example, when we examined the Wiley girls, we found that they had built out the top of two boxes, but had not started the bottom box. We could plainly see they were strong enough to take a third box at that time. I decided to do a different experiment there. We put the third box on top. This left them with, from top to bottom, empty, full, empty. I just want to see what they will do. Some beeks say, the bees always build up. Warré contends the bees want to build down, like they would in a tree. I suspect the real answer depends on their motivation for building: is it honey (up) or brood (down). Anyway, follow your instincts about what your bees are telling you. If they need another box, give it to them, top or bottom depending on what you see ... or feel ... while you are there.

Illustration 114: Hives for swarms are set up and ready in advance. The tall Lang is Mama Birch. All the others are empty, but ready with frames installed. Just to the right of Birch is two single deep hives stacked on top of each other. Just to the right of that is two more single deep hives stacked. These will all be filled shortly. Photo credit: David Braden

Moving and Merging

© steve kennedy 2014

During your first year as a beekeeper, you may not have any occasion to move or merge bees. Let's say your bees are in your own yard, carefully placed according to local ordinance and HOA rules. You only have one hive and it's a newly installed colony living inside. This doesn't seem likely to produce the need for a move or a merge. But, should you stick to beekeeping for a few years, the time will almost surely come when you do need this knowledge. You will have a hive in a place it is no longer needed or wanted. Those bees need to be moved. Or you will have a colony that is failing. You want the last of the bees to be productive for some colony rather than just fading away. Those bees need to be merged with another colony.

Some of us will have the need even in that first year. We set up a hive in a yard without having checked with all the neighbors. After all, everyone loves honeybees, right? But, the nicest neighbor comes knocking and explains that their child is deathly allergic to bee stings. Together, we decide to move the bees to a farm not far away. Or, you get a call from another first year beekeeper that has a failing colony, he has decided beekeeping is not for him and wants you to have the bees. The knowledge in this chapter will certainly come in handy.

Illustration 115: I know the door was here, really! No, it's right here! No, not over there. Photo credit: Don Studinski

Orientation

Bees orient down to the centimeter.

Guards that have reached the age of graduation to forager will perform a number of **orientation** flights to ensure they know where home is, and more specifically, where that entrance is. They program it into their very being. This is accomplished by flying increasingly larger circular patterns away from the entrance and then immediately back. Eventually, they will use landmarks to guide in from a distance, but the last little bit to the entrance is totally pre-programmed auto-pilot.

They know exactly where the hive entrance is and they will insist that this is true for some time even after it has changed. Want to test this? Just for fun, sometime when the hive entrance is pretty busy but still has an entrance reducer in place, about March around here, try rotating the entrance reducer 180-degrees. Now, instead of the entrance being slightly to the right of center, it will be slightly to the left of center. Observe the chaos that ensues. All the bees are confused. The entrance is no longer where they think it should be. They will figure it out, given enough time, but in the short run, things are a mess!

It is necessary to keep orientation in mind during many beekeeping chores, especially moving a hive. An old rule of thumb is that you can move *two feet or two miles*, but not in between. If you move a hive two feet or less, the bees will find the new entry point and go on with life without a large loss of foragers. It will be inconvenient for a while, but they will figure it out. Likewise, you can move a hive two miles or more and the foragers will "notice" that they are in a new location and initiate new orientation flights. But, moving an in-between distance may be asking for trouble.

There is an exception to this rule which I will describe later. Nevertheless, being sensitive to and knowledgeable about orientation instinct will help keep your bees alive and the superorganism together.

Moving Honeybees

Illustration 116: Taped and strapped, Superior ready to move. Photo credit: David Braden

The need to move a hive can come up any time during the year; however, it frequently comes up in spring. Not surprisingly, soon after swarm season starts, with splits just finishing up, the next hot topic many beekeepers are thinking about is moving a hive. Some split has created an added colony in a place it's not needed or wanted, therefore, it needs to be moved. Or some swarm has been placed in a temporary location while a permanent location is identified. Now it needs to be relocated.

Vocabulary:

Hive - a home where honeybees live, frequently, but not always, wooden boxes.

Colony - a superorganism consisting of honeybees that live in a hive.

Moving an empty hive is no big deal. Just try not to break your comb (it's fragile), or any wooden hive parts. Moving a living colony is another story. We want to save as many bees as possible, within reason, so there are several things to consider like temperature, forager location and distance.

Don't move bees in winter. Moving in cold weather is not recommended if you can avoid it. This is because the girls are clustered to preserve their heat. A bump or jolt will, at a minimum, disturb their cluster, and, at the worst, send them flying, which could kill them. The lowest safe temperature would be upper 40s°F. Below that, just be aware, you are taking a big chance. The move pictured here was done on a warm day in December in preparation for splits in spring. I needed the extra room next to Braden. Superior had to go elsewhere.

Because bees orient so specifically, moving a hive is not "necessarily" a trivial task. The beekeeper must be mindful of the fact that the foragers will go "home." That is, more precisely, where they remember home to be. Moving a hive 20 inches or so will probably not be deadly. They can, eventually, figure out where home is at that distance because they will smell the queen pheromone. Likewise, moving a hive two miles or more is not deadly because they will recognize that their new position is completely different and they will initiate new orientation flights. They notice that the sun and other orienting landmarks are in a different place with respect to their hive. This triggers their instinct to re-orient. They will begin with orientation flights which home in on where they are now located. Only when this work is completed can they begin foraging again. Between 20 inches and two miles is a different story.

Move bees at night while all the foragers are at home. That's right, think flash lights, head lamp maybe, and watch your step in the tall grass. Why do we move bees at night? It's because we want to keep as many bees alive as possible. The idea is to let all the foragers come home for the night. At that point, the entire superorganism is together inside the hive. The entrance can be completely blocked, trapping all the bees inside. Don't forget to provide ventilation. You can move them in this condition and every bee successfully arrives at the new location. If you move during the day, the foragers that are out will come home to a missing hive and they are doomed because they will stay there. Without a colony to support their life, they will die. At night, no bees are out flying.

Illustration 117: Just fit even with three deeps. Ready to secure. Photo Credit: Don Studinski

They are all inside the hive. You can approach the hive, turn the entrance reducer to fully closed and completely seal off the entrance with tape. Then you strap all the hive parts together with a tight strap (you don't want hive parts shifting during transit). I suggest using a ratchet strap. You lift the whole hive as a single unit. Bring it to its new location.

This is the part where you are really glad you chose to do this while they were in their first deep or even just a nuc. Once you have stacked them to three deeps and three supers, the very idea of moving them becomes unmanageable. This is also where having bees in more than one location can be an advantage. When you need to move them and you need an intermediate holding area (two miles away) it's nice to have a place for that intermediate stop all lined up in advance.

Here's a night-time moving procedure.

1. Make sure the new site is ready to receive the hive. Cinder blocks or pallets should be in place and leveled.
2. After all outside activity has stopped (dusk or later), approach the hive and turn the entrance

reducer to block all traffic.

3. Take a piece of duct tape and tape the entrance reducer in place. We want no bees exiting even if there is a bump or two along the way. Tape over any hole that provides an entrance/exit for your girls. This could be on deeps or supers that have been drilled for ventilation.

4. Get ratcheted tie-downs (one or two) and tie together all hive boxes, bottom board and lid into a single movable structure. This must be tight enough to prevent any movement between deeps or supers. However, not so tight as to break wooden ware.

5. You and a friend will be picking this up and bringing it to the move vehicle. Pick it up and put it in your truck, car or trailer.

6. Ratchet the hive in the vehicle to ensure there is no movement due to bumps, stops, starts, etc. Account for sudden stops. A spilled hive in a sedan could get really inconvenient.

7. Drive slowly trying to avoid a lot of bumps. This is especially true for frameless configurations like top-bar hives where the comb may break off the bar if bumped around too much.

8. Proceed to your new location. Of course, your base platform is already set up there and waiting for you.

9. Once you are on site at the new home, it should be as simple as placing the hive on the stand and removing the ratchet straps and tape. Keep in mind that you will have disturbed the bees somewhat and there will be some guards ready to come out and see who is there once the tape is removed. Be prepared accordingly.

10. Move the hive onto the new location platform.

11. Remove tape and ratchets.

Illustration 118: New home, straps and tape removed. Photo credit: Don Studinski

When the girls wake up in the morning, they will reorient and all will be well.

But, here's the rub. Let's say you have a hive on the east side of your yard and you really want to move them to the west side of your yard. Let's say it's 50 feet difference. There are two possible methods.

The first, I am absolutely dead sure will work. It's long and labor intensive. Move the hive two or more miles away. Wait a week or two to ensure they have completely settled in at the new site. Move the hive back to your yard and place it on the west side where desired. You have now successfully moved the hive from one side of your yard to the other.

The second, is a much simpler case, which I would also suggest you perform as a nighttime activity to save as many foragers as you can. Move the hive across the yard. Make a dramatic change to the "look" of the entrance. You are trying to get the foragers' attention. Every bee that exits must notice that something is very different. One idea is to place a leafy branch across their entrance so they have to find their way through the leaves to get out. Yesterday, it was a wide open entrance with no leaves. Today, it's a jungle. Hopefully, they notice the difference. If you don't find a bunch of bees clustered at the old location, then you have been successful. Otherwise, you may want to consider moving the hive back and trying the more labor intensive approach. I have had good luck using an evergreen bough across the entrance. This forces the bees to find their way through an obstacle course upon their next exit. They can't help but notice that something is significantly different.

Merging Honeybees

Combining two honeybee colonies, merging, can be smooth as glass or it can cause a war with a great deal of devastation to your honeybee population. Obviously, smooth is better for everyone.

A queen is a colony and a colony is a queen. The fact that you are contemplating a merge is an acknowledgment that you have a failing colony. They may be queenless unexpectedly or they may have a failing queen, but whatever the reason, you are about to lose a colony. The motivation to merge is to get the last bit of productive work out of a colony that will, eventually, fail. One could make the case that we are saving a colony when combining two small and weak colonies into one that can thrive. Instead of losing two, you keep one that can make it. But even in that case, you are still giving up on one queen and, therefore, you are losing one colony.

Then there's the case where the beekeeper planned for merge in advance: using two queens (two colonies) to quickly ramp up

Illustration 119: Top left to bottom right: Quay2, Araphoe, Cherokee, Pumpkin1.1. Photo credit: Susan Sommers

bee population, then performing the merge so that the large population colony can perform honey production. Large bee population colonies use more bees, proportionally, for honey production than do smaller colonies. Once the basic chores of home are covered, the remainder of the bees are free to forage. Therefore, having one home instead of two (after the merge) can produce more honey faster.

Whatever your motivation, there are three key factors to consider: queen, time of year and family.

Each honeybee colony will need to be limited to a single queen, eventually. This is because the queens will fight to the death, by instinct, under most circumstances. Ideally, we would choose a time to merge where the weaker colony, the colony with the queen that we want to eliminate, has a relatively small population. We need to go in and find the queen. That is made much simpler when there are 10,000 or 20,000 bees rather than when there are 50,000. Chances are that this will be the case. The small colony size is why you are motivated to merge. Search for the queen until you find her, no matter how long it takes. Pinch off her head. Now, this weaker colony is ready to merge with the stronger one using the newspaper method described below.

During swarm season, honeybees are easily manipulated to change queens. New queens are emerging, virgins are flying and workers are generally in the mood to be accepting of any queen. During this time, you can introduce a few pounds of bees to a queen, no problem, especially when many swarms are happening in the same area at the same time, like a large apiary. This is why packages work in spring. This is the time when you will see people shake nurse bees in front of a strange colony and they just walk in without difficulty. The rest of the year, honeybees are much more sensitive about who they will and will not allow entry.

Illustration 120: Crazy comb built upwards. Cherokee not looking strong enough to make it through winter. Photo credit: Susan Sommers

Family can also make a merge quick and easy. This is the case where you have tried a split and it failed. The bees in this newly split "colony" are the sisters of the mother colony from which they came. You can literally slam them back together without a war. They just pick up where they left off, one big happy family cooperating to maintain the life of the colony. Think of how happy the formerly split daughters are to have their mom back.

Newspaper merge is really quite simple as long as the equipment is the same dimensions. Consolidate the to-be-merged colony into a single hive box. Keep the best of the brood, honey and bee bread. Remove the top of the lower colony, including any inner cover. Place a single layer of newspaper over this colony; two layers at the most. Use your hive tool to slice a few little slits in the paper with the grain. You want a slice such that air can pass, but a bee cannot. Leaving some paper sticking out around the edges helps later when you want to remove it. At this point, you have a bottom board, a brood box and a layer of newspaper on top. Place the upper colony directly on top of the newspaper. Top off the stack with an inner cover and telescoping cover as appropriate. The upper colony needs an entry for foragers until the merge is completed. Many inner covers come with a slotted spot which can serve this purpose. If necessary, you may have to use a stick or something to provide a crack as an entry. The bees will eat their way directly through the newspaper. As this eating progresses, they become accustomed to the queen pheromone of the living queen. Queenless bees migrate into the queen right area and no one feels the need to fight because there is not a two queen loyalty issue. The living queen pheromone has been passed into the queenless area and even those workers will be considered sisters by the time they gain access to the queen right area.

This process can be reasonably successful even if you merge two living queens. You might choose to do this if you have two small colonies which need to be merged and you are not sure which is the superior queen. Nature will decide which queen lives. The end result is usually a queen right colony with a single living queen. In this case, at a minimum, the queens will have fought. One will be dead. However, I have seen this process fail. This implies that during the fight, both queens were either killed or adequately injured to render the entire colony queenless.

In any merge case, you will need to consider the issue of orientation for the foragers of the colony being moved. All the discussion of how to move a colony may apply. One way to influence this is to merge near dusk (so all foragers are home) and to not provide the top colony with an entrance. This will force the foragers to spend a day or two merging before they go out and increases the chances they will orient to the new location without a two mile move.

Here's an example. I knew from harvest inspection that Cherokee would not make it through winter. The queen never laid vigorously. Likewise, the second Quay colony had remained small and would not likely make it through winter without additional stores. These two happen to be at the same apiary which makes a merge pretty convenient. Let's merge Cherokee with Quay.

I ended up putting Cherokee on top of Quay. I checked Quay first and noticed a lot of comb built across frames. To get in and find the queen, I would have had to break a lot of comb and killed a lot of bees. I chose not to do this. Instead, I just left the top off in anticipation of placing Cherokee on top. I put the newspaper in place using the sticky burr comb to hold it while I worked. I checked Cherokee next and also noticed a lot of comb built across frames. Again, I decided not to dig in to find the queen. We chose not to find and kill either queen before performing the merge. We will let nature make that choice for us.

In this case, I sacrificed Cherokee foragers that were out because I wanted to perform the merge during the day as a learning event for students. Cherokee was between Arapahoe and Pumpkin. Quay was on the other side of Arapahoe. So, when I moved Cherokee to the top of Quay, it left an open space between Arapahoe and Pumpkin. Foragers, of course, came home to that empty space ... confused. Some drifted to Pumpkin. This induced some fighting on the front porch. Some drifted to Arapahoe. This induced more fighting on that front porch, noticeably more. So, in the end, I not only disturbed Quay and Cherokee, but also disturbed Arapahoe and Pumpkin. Probably not the best scenario, but such is life. Clearly, foragers are not welcome to enter an adjacent hive late in the season.

Since I have two Quay colonies and only one Cherokee, I'll choose to call the resulting colony Cherokee to eliminate the confusion.

The new Cherokee has a good chance to get through winter assuming they resolve the two queen situation without killing them both. I will not know until about 6 weeks after the merge. At that point, most of today's living workers will be dead or dying. If the population suddenly drops to zero, then I lost both queens. If not, a royalty female still lives.

In the end, Cherokee not only lived through winter 2012/2013, but she produced multiple daughter queens by splits in spring of 2013. It's true that by performing the merge of Quay and Cherokee, we sacrificed one colony. But, because they made it through winter and produced two daughter queens in spring, I am still one colony ahead. And, as of this writing, in January 2014, both Mama Cherokee and her daughter's colony are alive after some very bitter cold, well into negative territory. This is a case where I am sure glad I did the merge and I am sure glad I let nature choose the queen.

June ... Quiet Time

© steve kennedy 2014

J une is a great time of year to be in relationship with bees; nectar and pollen are readily available, so bees are busy and content, leaving them in a calm disposition.

Making observations at the hive entrance is the name of the game at this point. Saturday, I was out at Hearteye Village watching the Violet girls come and go. Perfect weather with comfortable temperature, a lot of sun and a light breeze. I find it easy to fall into meditation just staring at the honeybees. I lose track of time. It's wonderful.

Illustration 121: Taking notes. You can get into a zen state enjoying the apiary. Photo credit: Rennie Zapp

Swarm season is settling down. We are not getting calls every day anymore. Some recently installed swarms have not yet had their inspection to verify queen health. So, there are still some chores to be done. I have been relocating hives temporarily located at LSI. Just one left to move.

Launching into a chapter called Quiet Time, here's a fun photo of a time that might not be so quiet. Might I suggest some caution before moving that lawn mower.

I've been spending a lot of time lately thinking about 2013 splits. Beekeepers don't always do themselves a favor ordering bees from out of state. Along with the bees can come disease and parasites which may or may not be compatible with our local ecosystem. It would be best if we could satisfy the need for and demand for bees locally. Swarms represent good local genetic stock. The existence of a swarm tells us, beyond the shadow of a doubt, that a mother colony made it through our local winter conditions healthy enough to reproduce. Those are the genes we want to multiply for our local beekeepers. They represent genetic stock that can successfully negotiate our local disease, parasite and weather conditions. To multiply for the sake of our fellow local beekeepers will take some planning and preparation. Now is the time to start that planning and preparation for next year. There may be additional equipment purchases that have to be made and that may require financial planning, including saving ahead.

Illustration 122: Honeybees under lawn mower bag. Photo credit: unknown, public domain

I'm also thinking about 2012 honey harvest. We all need access to extraction equipment for either frames that can be spun or for crushed comb or both. It makes sense that we take steps to access common equipment rather than everyone satisfy this need on their own. There may be "per pound" fees involved, so don't let that shock you if it happens. So far, my main production hives, Birch and Superior, have NOT produced honey for harvest. This is agriculture and there are no guarantees.

Consider Your Varroa Load

Mites are not optional. Of 781 hives tested in 2011 across the USA, 92% had varroa[26]. Worryingly, 62% had varroa levels exceeding the damage threshold. Of the samples received in May of 2012, 100% tested positive for varroa.

Illustration 123: Sugar roll jar. Screen to fit by Norm Klapper. Photo credit: Don Studinski

Before we start discussing "managing" varroa, consider for a moment, the normal and natural living ecosystem within a hive. We know that for feral colonies, perhaps living within a tree hollow, there is a host of living organisms sharing the space harmoniously. At a macroscopic level we have mites, beetles, wax worms, ants and roaches. At a microscopic level we have fungi, bacteria and yeast. Some micro-fauna are actually necessary for pollen digestion in the bee gut.[27] This complex society of diverse

---

26   http://www.aphis.usda.gov/plant_health/plant_pest_info/honey_bees/downloads/2011_National_Survey_Report.pdf
27   http://bushfarms.com/beesmorethan.htm, visited 1/11/2014

organisms is performing countless symbiotic relational transactions, most of which we know nothing about. For example, *ascosphaera apis* causes chalkbrood but prevents European foulbrood. Is it a pathogen or an asset? How do you know? Do you really want to start messing with this complexity, which has proven itself to work for thousands or even millions of years, by introducing chemicals, essential oils or organic acids?

Expect to have mites and learn how to live with them. Colonies will die in their second or third year due to varroa mites. If you choose to keep bees chemical-free as I do, then there is one way to control mite population: break the brood cycle. Mites require honeybee brood on which to lay for their mite reproduction cycle. If you interrupt the availability of honeybee brood, then you interrupt the ability of mites to reproduce and thereby crash their population. There are many ways to approach this, but all involve removing an existing queen for quite some time and introducing a new queen. Whatever your chosen method, as long as you are able to force an extended time without honeybee larvae, you have broken the mite's ability to reproduce. My choice is to allow the bees to make their own new queen.

Consider testing for your mite load. There are nice procedures easily available on line[28]. I suggest you use a sugar roll rather than the alcohol roll because you don't have to kill the bees. Should your test indicate you don't have a mite problem, it may make a difference in how you proceed. On the other hand, if you find mites, as expected, then you will at least know the magnitude of your problem.

Let's say you have a large, healthy colony that is currently producing a good honey crop. Let's say you want them to stay on that path; you don't want to split them. But, you do want to interrupt the honeybee brood cycle. Well, the answer is to find and remove the queen. One way is to simply kill her. The preferred way is to capture her and carefully place her into a queen cage. Then you can use her to re-queen a failing colony or pass her to a friend. Within 4 hours, all the bees in the colony know they are queenless. They will begin emergency supersedure. They will make queen cells. Meanwhile, there is no laying within that colony. It takes 4 weeks for a colony to raise a new queen and get her to the point of laying. Meanwhile, the mites have no larvae on which to lay. The mites' need to lay is building up to a very intense level. Finally, when the new queen lays and the larvae appear, the mites will be in a panic. They will over lay on those first honeybee larva to appear. Their offspring will die due to insufficient food and a generation of mites is terminated. This interruption in the honeybee and mite reproduction cycles will set the mites back enough to allow the honeybees to regain the upper hand as a young queen lays vigorously.

The second case is "I want to save the genes of this colony, but chances are I will lose Mom." This is a split. Perform a normal split process. The new colony must raise a new queen. A queen takes 16 days to emerge. The colony will start with a young larva. The egg hatches on the third day. That larva is a viable queen for three more days. She will be capped on the 9th day. Therefore, the queen rearing process always starts between the 4th and 6th day and she will emerge somewhere between 10 and 12 days after the process starts. She needs a week to mature sexually, and another week to fly for mating. Therefore, new laying will begin in the 4th week after the queen rearing process begins. This new colony will experience the exact same interruption to mite reproduction as described in the first case. The mother colony, however, still has the old queen laying without interruption. There is no interruption to the mites. Therefore, we expect this mother colony to fade out in the fall as the mite population overtakes the honeybee population.

Of course, they might live. If they do, they are a very important colony which you really need to multiply in spring. If you don't know how to multiply or you don't want the daughter queens, then find

---

28 http://www.ent.uga.edu/bees/disorders/documents/VarroaMites_155.pdf, visited 1/11/2014

a friend who does. Your local ecosystem needs those bees. There should be beekeepers throughout your community that would be glad to receive the bees, and potentially, help you learn the split process as well.

In any case, when you are raising a new queen, you must check the colony one week after you have started the process. Start plus seven days means you are somewhere between 11 and 13 days along. Therefore, you are checking for capped queen cells (capped on the 9th day). If your colony has produced one or two queen cells, great, you are done for a few weeks. However, some colonies will have produced <u>many</u> queen cells. You need to terminate extras. I recommend not leaving more than four. Choose the biggest cells to keep. Also crush any partially built queen cells. If you fail to do this, the colony may swarm one or more times which can result in a depleted population which can be robbed, causing death. I once had a 50,000 bee colony produce 21 queen cells. They swarmed so many times that there were nearly no bees left. Neighboring hives promptly robbed their honey. They were dead in two weeks. What we are trying here is an emergency supersedure. We don't want them to swarm at all. Of course, there is no guarantee that they will not swarm. However, if you have provided an appropriate amount of room for expanding brood, then you increase the chance of successful supersedure without swarm.

You will want to perform mite control splits soon <u>after</u> the summer solstice, June 21, in the years it is appropriate for you. You might not get to it until July, but if you delay too long, then the bees will not be able to adequately prepare for winter.

Use June to catch up on some rest. Get some reading done. Your supers are on, right? Make a wish list of new equipment and begin saving for the stuff you will purchase. Will you be building some? Dream of honey. It's not far away.

Illustration 124: Smoker open. Photo credit: Nancy Griffith

I am considering requeening Birch or Superior or both. They are 2nd and 3rd year colonies. I have not yet made a decision. I'm disappointed that I have no honey and I'm hesitant to interrupt the brood cycle. There's a good chance that, if I choose to split, I will get no honey harvest in 2012. I may prefer to allow the colonies to die giving me honey. That's not entirely bad for me because it will free up equipment that I will need for splits in early 2013: deeps, brood frames, drawn comb, honey and bee bread. The decisions have not yet been made.

Smoke Alternatives

Not a hot issue this time of year, smoke will become of much more interest after solstice. It's not like throwing a switch, but after solstice the bees will begin getting increasingly defensive about their stores. Therefore, our likelihood of encountering a hostile response grows more with every passing day after the daylight begins to wane. By late summer and into fall, smoke will become a necessity.

I like fiber fuel made of compressed cotton fiber. It lights (comparatively) easily and stays lit for the duration. You will learn the right sized chunk over time and can judge how much fuel you will need for the amount of time you plan to spend with the hive open. You can find this fuel on the web and have it delivered to your home. Sometimes it arrives in very compressed cylinders which can be hard to break apart, but the burn lasts longer. Other times it arrives in lightly compressed cylinders which lights very quickly, but doesn't last as long.

I frequently use sticks, leaves and grass that I can easily find at the apiary to supplement the compressed cotton. When I'm there for an extended time, I might start with the cotton chunk and make adjustments with sticks and grass while I'm there. If you are into recycling and try to find a use for every scrap, like me, you can save citrus peels and dry them out. They work great stacked on top of the cotton chunk and give the smoke a pleasant aroma.

While I do think misting the bees with sugar water or water combined with various essential oils, like lavender, may have a short-term calming effect, it is my experience that depending on these techniques 100% of the time would be an act of folly. These techniques may work fine in the early part of the year from mid March to solstice when the bees are generally calm and they may be great for a very short peek in summer. But I would certainly not recommend it for a testy colony, or for any colony in fall.

Honeybees are mostly calm, except for when they are not. Weather changes, barometric changes, things which may not be obvious to us, can cause a large disposition change in bees. It is their job (at times) to be defensive, that's partly why they are still on the planet after 18 million years. A judicial puff of smoke, unequivocally has a calming effect. With that being said, I understand the difficulty of only having moderate strength in the hands and coordination challenges which can make the smoker difficult. Sugar water can also present inconvenient difficulties when the sprayer's spout gets clogged or the darn thing leaks all over your tools.

Probably the most common complaint is that the smoker is hard to light and then it's hard to keep lit. Try getting a propane torch. Fill your smoker with the fuel of your choice. Now light the torch and let the flame hit the outside of your smoker below some of the fuel. The intense heat will go right through the metal and make the fuel ignite. Give it 30 seconds or even a minute or two. You will have a nice set of embers started. For the duration of your work, don't forget that your smoker needs oxygen or it will go out. You can't get lost in your thoughts about bees and forget to puff the smoker. But, you can set it down from time-to-time and pop the lid open. Set it up-wind and let it drift the smoke across your work. The bees don't need a lot of smoke. Just the cool drifting smoke is all you will need most of the time.

Here's another idea for those that are smoker challenged. I've seen video of an old Italian gentleman who had a small smoker that hung around his neck. It was in a small brace, much like what a guitarist would use to hold a harmonica. To operate, one just had to blow through it--leaving the hands free to work. I haven't seen one in years, but I am sure they must be available somewhere.

Mite Control Split Planning

Choosing to perform a mite control split requires the beekeeper to get their plan in place in June so that the work can be performed in late June or early July. I have been thinking over my options and have decided not to split either Birch or Superior. I want that honey. I've spent years getting those two colonies ready to produce and I want my payback.

I have two purchased colonies out at Rock Creek Farm. I call them Pumpkin 1 and Pumpkin 2. I don't know much about them other than that they survived winter of 2011/2012 and seem to be quite healthy now. So, I figure they have varroa mites. I want to use them as an experiment. I'm going to split one and not split the other. Then, we will see who is alive in spring of 2013 and who is dead.

In this case, we will perform a box split. We will not know where the queen is until seven days after the split when we go looking for queen cells. Not related to the split, but combining chores where possible, the hive will be disassembled to the hive stand. Adjustments will be made to face south, level and make room for two hives side by side. The hive cannot be moved very far, but needs to move some. We will take the three deep hive, Pumpkin 1, and make two hives, one with one deep and the other with two. The two deep hive will keep the super. The one deep hive will probably be the middle deep. I'm hoping the old queen will be there. The newly configured hives will sit side by side with returning foragers not knowing which to enter. They will eventually figure it out.

This will be my only post-solstice split this year. It's about living with mites.

Illustration 125: Rock Creek Apiary. Back to front: Pumpkin 2, Cherokee (left), Arapahoe, empty, Pumpkin 1.1 (right), Pumpkin 1. This is just before the cow got in and made a mess of everything. Photo credit: Don Studinski

The way this played out is that Pumpkin 2 died in the fall, probably of mite overload. In December, a cow gained access to the apiary and knocked over several hives, including Pumpkin 1, which died shortly afterward. The daughter queen, Pumpkin 1.1, made it to spring, but died in a vicious cold in April of 2013.

Honeybee Tummies

Record heat with several days over 100-degree F has me thinking a lot about my bees and their needs. This heat may shut down a lot of flowers. Not to mention we have not had rain in a long time.

Bees must have comb. Without comb there is nowhere to lay eggs, nowhere to store nectar and nowhere to store bee bread. These are all necessary things for colony survival. To build comb, the bees must have full tummies. Without adequate nourishment, their wax glands will not function.

Honeybees have a **crop**. It's the first stop for food after passing through the esophagus. The crop is sometimes referred to as the **honey stomach**. The crop is the structure used to carry nectar from flowers to the hive. A typical nectar load is about 20-40 mg, but this organ can hold up to 100 mg. One pound of nectar requires 12,000 to 24,000 foraging trips. One pound of honey represents many more trips because nectar is 80% water. The **proventriculus valve** prevents nectar from passing into the stomach (**ventriculus**). The individual bee is nourished by food that is allowed to pass through to the stomach. Nectar not allowed to pass through can be easily expelled to allow for further refinement into honey.

One factor affecting full tummies is the current nectar flow in the immediate geographic area. Another is how much food the bees brought with them. What does "brought with them" mean? Well, potential sources of honeybees include nucs (most expensive), packages (next most expensive), splits (free for yourself, otherwise, maybe not), swarms (frequently free, but not always) and removal (someone pays you to get these bees out of a structure). Each of these sources results in a different mix of honeybees in terms of age, capabilities and tummy content. Tummy content is what they brought with them.

A nuc is really just a small colony, fully functioning, with an actively laying queen. The term "nuc" is also used to refer to the box in which they are living. But here, we are referring to the colony. All age groups are represented. All jobs are filled. Comb has been drawn. Brood is present as is honey and bee bread. Pheromone levels are in balance. Maybe that's why they are so expensive. These bees have full tummies from a normal cycle of eat and cleanse. Whatever comb they need to build, they should be fully capable, as long as they have access to nectar and pollen.

Packages are filled with unrelated workers from any number of colonies that got dumped together. They have been forced to be queenless for some time. A completely unrelated queen, in cage, has been introduced and everyone is expected to get friendly. It's much more stressful on the bees compared to a nuc. Another complicating factor is that the workers were not given any opportunity to prepare for this ordeal. They must be fed during transit or they will die. After installation, I do not feed a package, assuming the flow is on. However, if my geographic region is in dearth, then I

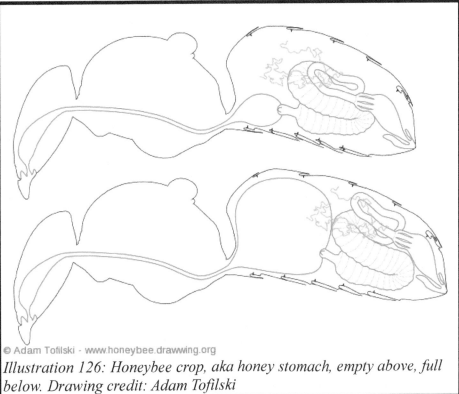

© Adam Tofilski - www.honeybee.drawwing.org

Illustration 126: Honeybee crop, aka honey stomach, empty above, full below. Drawing credit: Adam Tofilski

would feed. These ladies are potentially hungry. It takes full tummies to produce comb. In many cases, you will expect your package to produce new comb. The more you expect, the more you should consider feeding.

A split is more like a nuc than a package; it just started at your home apiary rather than somewhere else. They should already have all they need.

Swarm is a special case. These ladies are engorged on honey before they fly. They are fully prepared to build comb upon arrival. Further, it's the young bees that fly because they are the ones that make comb. Are they docile because they have no guards or are the guards just friendly because they have no home to defend?

Removal, or cut out, is both the best case and the worst case. It's great when someone is willing to pay you to get bees. What a deal. Unfortunately, colony survival is relatively low. Removals frequently involve some pretty brutal treatment of the insects. There is death and destruction. It's easy to hurt or kill a queen by mistake. The bees, of course, have had no warning and have made no preparations. It's likely most or all of their brood are damaged or dead due to exposure. Will they be able to make comb? This is most likely a case where you must feed to help them along. If they need to build comb, they need full tummies.

Which brings me all the way back to the thinking that got this started. I'm going to be watching closely for colonies that are consuming their stores due to this brutal heat and will, therefore, need to be fed. If this weather keeps up, then I may have to break out the feeders. By reading this carefully and thinking about the case where your colony fits in, you will be able to determine if you need to be feeding.

Student Questions

Susan asks:
Why is it that varroa control splits have to occur after the summer solstice?

Answer:
We want to interrupt the mite reproduction cycle late enough in the year that the mites do not have time to recover before the bees go into winter while still being early enough that the bees have time to raise a new queen and she has time to raise a good winter brood. After the changing of the days (a little less light each day from solstice forward), the bees start changing their strategy to account for different nectar and pollen flows and winter preparations. From solstice forward, you will start noticing an increasing defensiveness when you inspect your bees. They are very tuned in to the annual cycles even though their lifetime is not long enough to live through a full cycle (except the queen).

Illustration 127: Bearding. Photo credit: sourcherryfarm.com

Linda asks:
What is **bearding** in bees?

Answer:
Bearding is a temperature control strategy. Remember when your Mom got fed up with your noise and said, "Go outside and play." It had nothing to do with you "wanting" to go outside. It had to do with Mom "needing" you to go outside. Bearding is like that with honeybees. Bearding is when a bunch of bees go out on the front porch. Sometimes so many go out that they have to hang together looking like a swarm cluster. It also looks somewhat like the hive itself has a beard hanging on the front. Getting those bodies outside is helping reduce the temperature of the inside. You might want to double check that they have adequate

ventilation, but it's probably nothing to worry about.

Chris writes:

Over the recent hot days, I have noticed some interesting bearding activity between my hives, which makes me think the Warré is better at managing temperatures. Let me explain.

Both my Langstroth and my Warré get significant late day sunshine - usually our hottest part of the day. The Warré girls will beard, not heavily, between 3 & 5 PM. As dusk approaches, the Warré girls are still flying around, but there is virtually no bearding.

The Lang girls start bearding, heavily, after 5 PM, and will continue well into the evening and after dark. The big, obvious difference between the hives (other than hive type), the Lang is very full. The Warré is currently three hive bodies; bottom box added yesterday, and therefore, empty. Top hive box is very full of bees, honey, wax, brood, etc and weighs about 35 lbs. Second hive body is just being built out with comb (appears straighter than top box!).

So these observations lead me to think the Warré is better at helping the bees manage the summer temperatures.

What are your thoughts, observations at your hives?

Answer:
What you suggest may be true. I am not sure. Here's another possibility: Because the Lang is so full of bees, bee bread and honey ... all heat absorbing matter ... maybe it is more of a heat retainer than is the Warré, thus driving the bees out on the porch. Likewise, even though the Warré has a heavy top box, maybe because they have so much free room heat is more easily dissipated.

Bearding is a function of heat and ventilation. Inside temperature is affected by the mass within the hive. Some combination of heat being absorbed and not dissipated by ventilation will drive the bees out. I've seen this happen on every hive type. Therefore, it's probably not

Illustration 128: Top-bar hive bearding. Photo credit: Brook Franz

specifically related to hive type so much as it is what the bees have built inside and how they are managing the air flow.

Leslie asks:

As I was explaining the split to my friend, he asked why we didn't kill or remove the queen so that both hives would have to breed a new queen, interrupt the mite cycle and survive. It's a valid question. You seem to practice as little interference as possible but saving a hive sounds worth it.

Answer:
This is a valid question I have been asked by several people. The answer may be disappointing. It's

because I'm lazy. The queen is one bee among about 50,000, maybe more, on at least 20 frames, maybe 30 frames, maybe more if I'm not using a queen excluder. I'm specifically trying to avoid having to find the queen. It's long, hot, sometimes frustrating work. Earlier this year, Linda and I had to find the queen for a split. It was a different scenario. That took us over an hour. We had to pour over 20 frames twice. So, yes, it may be worth it, depending on how hard you are willing to work to find that queen. Also, if you kill her and there is no viable egg or larva available, then the whole project fails. On the other hand, if you attempt a split and you don't kill her, but the split fails (no viable egg or larva), then you can always just combine them back. They are sisters and will combine immediately.

Another factor is that sometimes that queen will make it, even though we thought she would not. In that case, I increasingly want daughter queens from her because she has proven additional value.

Brian asks:
I have a lot of wasps drinking from the bees' water. Is this a problem- other than me not being able to get too close to refill the water anymore (I'm spraying from the hose)? I wasn't sure if the wasps chase the bees away.

Illustration 129: Watering Station. Photo credit: Linda Chumbley

Answer:
The bees are fine among the wasps. The wasps may actually be helping to keep the wax moth population in check. I don't even think the wasps will bother you if you approach the water. It's the

water they seek. We need all living creatures. I would let them be.

Bobbi writes:
I did see a few dead bees on the entrance one morning and then they were gone by afternoon. Do bees leave the hive to die or do they drop dead anywhere - including inside? I had read that a healthy hive pushes the dead-out.

Answer:
If at all possible, bees leave the hive to die. They know that hauling out a body is a lot of work for their sisters. However, there are numerous reasons that flying away may not be possible. Therefore, you will frequently see undertakers dragging bodies out. Sometimes bodies are dumped at the entrance. Sometimes they will fly the body away for a few feet or a few yards.

Superorganism

© steve kennedy 2014

Ahoneybee colony is considered to be a superorganism. A superorganism is an entity, like a colony, made up of many distinct individuals, that has the ability to act on the environment to achieve specific goals using highly specialized division of labor.[29] The individuals that make up a superorganism cannot survive for extended periods independent of the superorganism. And, of course, there is no superorganism without the individuals.

Illustration 130: Lafayette, Colorado swarm. Photo credit: Molly Turner

The existence of superorganisms in the insect world, like honeybees, is fairly common knowledge. Not so widely

Illustration 131: Naked mole rat. Photo credit: public domain, from Wikipedia

known is that there are also superorganisms in the mammal world, for example, the *naked mole rat*. Some scientists suggest that the human body is an example of a superorganism as they consider the fact that our digestive tract, is made of 10^13 (that's 10 to the power of 13) microorganisms. Other superorganism examples include ants, termites and coral. Superorganisms are made of a tightly knit collective of individuals, formed by altruistic cooperation, complex communication, and division of labor.

A superorganism is governed by the collective to accomplish goals for the good of the collective. In the world of honeybees, we call this the hive mind. But, the honeybee hive is made of more than just the queen, workers and drones. Beehives also house complex communities that bind together into a hive-centered ecosystem including other insects, microbes, viruses, bacteria and fungi. Some of these are even further specialized into specific areas of the hive interior, for example, the pupae, stored bee bread or honey[30]. All these organisms and food sources interact to maintain a balance of eaters and eaten.

James Lovelock and others have postulated that the complex system making up our entire biosphere is also a superorganism. In that case, everything, including us, is participating in a hive mind. From the Preface, page ix, of

Illustration 132: Blue Marble. Photo credit: public domain, NASA's Earth Observatory

---

29 http://en.wikipedia.org/wiki/Superorganism, visited 1/13/2014

Gaia, A new Look at Life on Earth[31], by James E. Lovelock[32]:

> Journeys into space did more than present the Earth in a new perspective. They also sent back information about its atmosphere and its surface which provided a new insight into the interactions between the living and the inorganic parts of the planet. From this has arisen the hypothesis, the model, in which the Earth's living matter, air, oceans, and land surface form a complex system which can be seen as a single organism and which has the capacity to keep our planet a fit place for life. [underline added]

Another term for hive mind is distributed intelligence. This term reflects the fact that many individuals with limited intelligence and information can pool resources to accomplish that which is beyond the capabilities of any individual within the collective.

Life is much more complex than most humans acknowledge. In fact, it's so complex that it is neither understood nor understandable. That's why we should not alter life's path by distributing chemicals across Earth's surface. There are unforeseen consequences, always.

Illustration 133: Smallmouth Bass. Photo credit: public domain, from Wikipedia

Take just one small example of a complex, synergistic, system. Smallmouth bass contain a parasite called yellow grub[33]. Yellow grub eggs are passed into the water by fish-eating birds where they are picked up by the fish. The more parasites a fish contains, the more likely it is to swim near the water surface where, consequently, it is more likely to be eaten by a fish-eating bird. Inside the bird, the parasite lays its eggs and the cycle begins again. Without the bird, there is no parasite. Without the parasite, there is no fish. Without the fish, there is no bird. What came first? What comes next? When we expand this concept to all the species covering the earth, we see how this system is complex beyond human understanding.

Each honeybee has an individual existence. Yet, the life of the colony comes from the way in which the honeybees work together. Below some threshold of honeybee population, the colony cannot survive. The superorganism breaks down; it dies. We don't know what that threshold is and it may be unique from colony to colony. This may be affected by climate and individual honeybee health as well as other factors.

As David Braden, Principle of Living Systems Institute, says, "You are already participating in a set of interactions involving all the living things around you. There really is no boundary to 'around you'. Your community ultimately extends to all things. Every action, every choice, of every living thing, has repercussions throughout the community. We generally think of our community smaller, all those living

30 http://grow.cals.wisc.edu/featured/protecting-our-pollinators/, visited 1/13/2014
31 http://en.wikipedia.org/wiki/Gaia_hypothesis, visited 1/13/2014
32 http://en.wikipedia.org/wiki/James_Lovelock, visited 1/13/2014
33 http://en.wikipedia.org/wiki/Clinostomum_marginatum, visited 1/13/2014

things within the range of our ability to influence their interactions, and we call that our habitat."

Within our habitat, we are an individual contributing to the hive mind of our superorganism, also known as us.

Illustration 134: Honeybee Keep logo. Photo credit Betsy Seeton

July ... Mite Control

© steve kennedy 2014

Having planned our post-solstice splits in June, we are now ready to execute that plan. Any equipment moves that needed to be made in advance should be done by now. The colony candidates for post-solstice split include second or third year queens which are likely to be carrying a heavy mite load. Performing a mite count prior to performing the split will increase your confidence that you are on the right track, and it holds the possibility to change your decision if the mites are not there.

The actual split process is no different than it was in spring when we were splitting to increase our colony count. The only difference here is the motivation behind the split.

Better Have Smoke

Now that solstice has past and the days are growing shorter, your bees will grow increasingly defensive about their stores. If you have not yet learned to use that smoker effectively, then you will soon be encouraged to do so. Here's an example of bees in the air. This is an example of a major intrusion without using smoke. This is a split. Notice the date on the photo. Despite the fact that this is well before solstice, the intrusion has them quite upset. After solstice it would be worse.

Illustration 135: No smoke. Bees in the air. Photo credit: Marci Heiser

This is another example where every frame was inspected using smoke. Take note of the difference in the number of bees in the air. This is the same time of year as the previous photo. Be sure to take note of the smoke at the bottom of the photo. All it needs to do is drift through the area.

Illustration 136: Inspect using smoke. Photo credit: Molly Turner

As the year progresses, especially after solstice, the bees will get increasingly more defensive. Here's an example of a newbie that is finding this out the hard way.

> Hi, everybody,
> Well, I've been using this veil to work my hives and before, when there were less bees and no honey, they didn't seem to mind me and all was okay. But now, I've been cleaning up a top-bar hive and this disturbs them. They've been getting in my veil and my hair which is no fun! So, I'm in the market for a new veil. Any suggestions?

If this person would take the time to learn to effectively use a smoker, the problem would be solved.

Meanwhile, it's too hot to be wearing a jacket to inspect a hive. Luckily, we have done all the preparation for this time of year. We can enjoy watching the front door and leave them alone. Next check is Pumpkin 1.1 on 7/27 or 7/28 ... what will we be looking for and why? Will it be necessary or advisable to use smoke? Why or why not?

Bill says:

I'll say it would be advisable to use smoke, its now late in the season, fall/winter approaching and bees will be very protective of the hive.

This is exactly right. Honeybees, quite gentle before solstice, may have a much different attitude after solstice. Think about what this tells you about photos you see with beekeepers wearing no protection. What time of year did they take those photos? Probably spring, maybe summer, but probably not fall.

Varroa destructor

Our girls, *Apis mellifera*, suffer greatly at the hand (or "mouth") of varroa destructor. This is the mite blamed for much of the destruction among honeybee colonies in recent years.

Illustration 137: Varroa mites on pupa. Photo credit: public domain from wikipedia

Illustration 138: Apis Cerana queen and workers. Photo credit: public domain, from www.researchgate.net/post/Asian_ honeybee_Apis_cerana_apiculture _status_and_bottlenecks

We seek honeybees that can live successfully in the presence of varroa mites. We know this can be done for a variety of reasons (consider Varroa Sensitive Hygiene (VSH) and Minnesota Hygienic honeybees), but the most important evidence is *Apis cerana*, the Asian honeybee, whose natural defenses against the mite allow cerana to co-exist without negative effects[34]. To increase our chance of getting some mellifera honeybees to live long enough to evolve the natural defenses we seek, we should understand the reproductive cycle of varroa, side by side with the reproductive cycle of mellifera. Taking advantage of that knowledge can give us an edge in choosing a management technique.

Credit is deserved by Mel Disselkoen whose website[35] got me thinking about these issues several years ago.

There is one key issue of particular interest here: varroa cannot reproduce without access to honeybee larvae.[36]

Varroa mites are ectoparasites that feed on the hemolymph (honeybee blood) of immature and adult honeybees by puncturing the soft tissue between abdominal segments. The mites also act as vectors for disease, like deformed wing virus (DWV), to enter the honeybee. The reddish or dark brown mites (female) are oval and flattened, but they are visible to the naked human eye, roughly the size of a pinhead. This is an advantage to the beekeeper when analyzing mite load within a colony; dead mites can be seen on the bottom board with close inspection. Mites

Illustration 139: Drones at the front door. No challenge. Photo credit: Ken Thomas, public domain, from wikipedia

fit into the abdominal folds of the honeybee, holding position using setae (bristles) and avoiding the

---

34 http://www.beesource.com/resources/usda/the-different-types-of-honey-bees/, visited 8/15/2014
35 http://www.mdasplitter.com/index.php, visited 8/15/2014
36 http://www.biosecurity.govt.nz/pests/varroa, visited 8/15/2014

bee's natural grooming. Male mites are smaller, yellow and spherical in shape.[37]

Honeybee drones are known to "drift" from hive to hive and are generally allowed to enter without guard challenge. Varroa take advantage of this to spread their population. Adult female varroa enter a honeybee hive by hitching a ride upon a drone. Less frequently, this can also happen via worker bees which will generally be greeted by a guard challenge at the front door of a hive not their own.

The mite will leave its transport bee to enter a ready-to-be-capped brood cell. Once this is done, the mite is now known as a **foundress** mite. They prefer drones. In the cell, they crawl under the larva and hide in the brood food waiting for the cell to be capped. They then feed on the defenseless prepupa, and, finally, lay eggs. Mite eggs are white, so small they require magnification to see. The female places the eggs on the cell wall. The first egg laid, toward the cap, will be infertile and will become a male mite. Subsequent eggs will be laid every 25 or 30 hours, toward the midrib, all fertilized female eggs, generally four or five. Once hatched, the mites go through two juvenile stages (protonymph and deuteronymph) and then become adults in six or seven days (both sexes). The mites complete their mating while in the capped cell, brother with sister, unless multiple foundresses entered the same cell. When multiple foundresses enter the same cell, too many eggs get laid for the accessible food. The pupa will probably not survive and the juvenile mites may starve. (This is important for later).

Varroa cannot reproduce without honeybee larvae. The juvenile stages only happen in the honeybee larva's cell. The male mites will never leave the cell, therefore, all discussion of the mites outside the honeybee cell should assume a female mite. Females leave the cell riding their host. Although she is capable of laying eggs immediately after she matures, she is more prolific if she takes a phoretic period where she rides adult honeybees eating hemolymph. The phoretic period lasts 4.5 to 11 days when honeybee brood is present and may last as long as 6 months, in the absence of brood, allowing the mite to overwinter. Seven days to grow into adulthood plus 11 days to ride gives us 18 days of mite life thus far, that's the maximum of the range, the worst case.[38]

Female life expectancy in the presence of brood is 27 days, but in the absence of brood, many months. That second point, the many months part, complicates things considerably. If you want to pick on this theory, concentrate on the many months part. It completely ruins the math on which rests this entire approach. For the sake of this strategy, we happily note the 9 days that round out her life from the end of the phoretic period to the end of life. We zero in on that 27 day total.

Given mating failure and mortality in the cell, female mites can produce two viable female offspring for every drone larva used. When worker larva is used, only one viable female is produced. If she is only replacing herself, one for one, that case is not of concern. Therefore, I'm ignoring the worker larva case because it does not cause a growing mite population.

Let's assume a female life of 27 days and she produces two to replace herself during that time, then dies. Further assume that this happens every 27 days for 180 days (6 months), six and two thirds cycles. We will round down to just six cycles. That's $2^6 = 64$. One mite becomes 64 mites in six months. Note the six is the number of cycles, not the number of months. It's $2^{14} = 16,384$ if year-round (365 / 27). That's why mites can quickly overcome a honeybee population and kill their host.

Varroa have traversed the world and are found nearly everywhere today. Here, in the USA, they are always present. Always assume mites are there until you have proven otherwise. In fact, the National Survey Report shows that 100% of colonies tested in May of 2012 had varroa mites. Prevalence for

37 http://en.wikipedia.org/wiki/Varroa_destructor, visited 8/15/2014
38 http://entomology.ifas.ufl.edu/creatures/misc/bees/varroa_mite.htm, visited 8/15/2014

2011 was nearly 92% in 781 samples and 62% of those were over colony damage threshold[39]. You have mites whether you realize it or not. They are multiplying. What should you do?

Here's what I do with some success. Am I eliminating mites? No. I'm learning to live with them.

I assume a <u>first year colony</u> will not produce a honey crop, but will produce enough stores to get through winter. A first year colony, from a split or a swarm, which successfully <u>lives</u> through their first winter, is a <u>candidate</u> in spring of the next year for a split. That is, a 2012 swarm that lives, is a candidate for a split in 2013. The motivation for such a split is increasing colony count and may cause all resulting colonies to produce no harvestable honey. A first year colony that <u>fails to live</u> through winter (dead before spring 2013) represents genes I want <u>out</u> of my pool. I smile and say, "Yippee, that equipment is now freed up for better bees."

I assume a <u>second year colony</u> is carrying a mite load that will kill it come fall. I want some of my second year colonies to produce a honey crop. Those that do are showing me genes (honey harvest and alive) I want to keep. Those that don't (just alive) are not as attractive. That harvestable honey will be put up by summer solstice, usually June 21. After solstice, colonies showing genes I want to preserve <u>shall be split</u>. I want those daughter queens reigning over colonies ready for the coming winter. I <u>expect to lose</u> the mother colony. I'm counting on the daughter queens to carry on the line. The motivation for summer splits is not colony count increase, but rather is saving the line despite mites. It's about living with mites, not killing them.

My good friend and mentor, Eric Smith, points out that the varroa that kill their host colony are self eliminating. They are removing themselves from the varroa gene pool. This is to our advantage <u>and</u> the varroa advantage. If they don't kill their host, they will be more successful long term.

If the mother colony <u>dies</u> as expected, then there may be honey that can be passed on to another colony or harvested for humans. Passing on honey should take into account the varroa life cycle before the honey is delivered. Wait 30 days or more and you are assured that the mites that were living on that frame are now dead. If the mother colony <u>lives</u>, then I smile and say, "Good for you." That next spring, I will want daughter queens all the more. That queen is showing genes that can live with mites.

Splits in spring are about increasing colony count. Spring splits are focused on colonies having lived through their first winter. Splits in summer are about living with mites. Summer splits are focused on colonies having lived through one or more winters. The motivation affects the timing and which colonies are manipulated.

When I split, in any case, I make the colony create a new queen from scratch. This is nature's way, but more importantly, it requires the time necessary to knock down the mite population. I'm hoping to take 27 days or more, thus exceeding the expected lifetime of that female mite. I want a new mite, egg just laid on a drone on split day, to be dead before the next larva is available from the new queen for a female mite to use for reproduction. If I can do this, it will cause a crash in the mite population. A major setback in favor of my bees. Here's how it plays out. We will always assume the fast track to the next larva available for a mite. That's the worst case.

Day one, split day, colony is queenless with mites. They pick a candidate larva to become a queen. That larva cannot be older than 6 days or it's not a candidate. She will be capped on the 9th day. She will emerge on the 16th day, just one week after being capped. At this point we are 10 days after split day (or split day plus ten days). It could take longer if they choose a younger larva to become queen, but it can't be done faster. No new honeybee eggs have been laid during that time, therefore, no larvae are available to mites.

Digression: There is a hole in this strategy. Think about the drone eggs that were laid on split day just before the split, they will be "ready" for the mites on day 11 (tomorrow). This means mite reproduction continues up to the 11th day after split day. We will touch on this again later.

Back to the main line of thinking. The new queen must mature about one week prior to nuptial flight. That puts us 17 days out from split day. Fertile mites are getting anxious because they have no larvae on which to reproduce.

Illustration 140: Queen Cell. Photo credit: public domain from wikipedia

The new queen spends about a week flying, but we will assume that she only spends 5 days. Now we are at 22 days from split day. Mites alive on split day are approaching natural end of life.

The new queen comes home and starts to lay. For this discussion, we say she begins immediately (this may not be the case). Those eggs will remain eggs for three days, then hatch and become larvae. Now we are 25 days out from split day.

Mites are not attracted to a larva until it's ready to be capped, which for a drone is the 11th day, that's eight days after it was an egg. This brings us to 33 days since split day. Even if the mites choose to jump in with a worker, we're still at 31 days. There are still mites in the colony. Remember that 11 days up front that drone larva might be available. Furthermore, mites may have just arrived riding on a drone. But, the population that was already living in this colony before the split has been nearly eliminated. Most of the reproductive-capable mites crossed their 27 day life cycle and have perished.

Those still living are in a panic to enter a cell and lay. Multiple foundresses in single cells are likely because there is a small amount of candidate larvae available. This will, in turn, likely kill the larvae due to mites overlaying in a single cell and, thus, kill the mite young by starvation.

We killed the majority of the adult mites by waiting for their natural lifetime to end. We killed the next generation of mites by forcing a situation where the remaining adults overlay on the few honeybee larvae laid. This gives my new colony a nice fresh start. They will be ready for winter with low mite count. We're able to do all of this because varroa must have honeybee larvae. Manipulate the honeybee larvae and you can manipulate the mites.

This thinking is not about split technique or procedure. It's about the <u>day count</u> and how it can be used to help honeybees live with mites. Such an approach is saving money by not participating in chemical purchases and expanding the gene pool by creating more queens with more combinations of genes.

This technique helps nature produce the correct genes to co-exist with no negative effect. Compare this to the alternative of putting on miticide (acaricide) and hoping the colony makes it through winter without performing the split. In that case, we add to the poison in the environment and have not expanded the gene pool. The splitting technique is also about buying time; time to allow nature to do what nature does best: create resilience through diversity. By knocking down the mite population temporarily, I give a colony another opportunity to reproduce. More queens creates more diversity. No poison was needed.

Student Questions

This year, Bill got a colony from a removal which was taken the day after it swarmed. This sort of makes it a swarm. They swarmed, moved into the vent area, got removed and moved to Bill's all in 48 hours.

Illustration 141: Queen Street Removal. They entered behind the vent on the left. Photo credit: Don Studinski

Illustration 142: BeeHaus frame. Different dimensions from all other hives. Photo credit Bill Koeppen

It's a good candidate for a 2013 split if it lives through the winter. It got placed in a Beehaus which is not compatible with <u>any</u> of my existing equipment. So, Bill kindly offered to design and construct some modified frames which will facilitate the split next year.

Equipment Customization

Bill writes:

I'll have time to create the modified frames for the Beehaus and your swarm, what kind of a time frame are we looking at - if the hive is strong and progressing well when would we need to place the frames? Or if the heat has had adverse effects is it more likely we'll need to focus on keeping the hive healthy and alive through the winter?

Answer:
Please proceed with creating the frames. We will need to introduce frames slowly, one at a time, in the brood nest. We should

Illustration 143: They are about 12 feet back hanging between floor joists. Photo credit: Don Studinski

start now. We can only do this in warm weather because we will be separating the brood nest into two pieces. In cold weather, when they need to cluster, two brood nests is very difficult on them. We are trying to force them to build comb and lay on these frames. That's why we are putting them in the middle of the brood nest. We will introduce one new frame, then wait two weeks and check their progress. If that one is being used, then we introduce the next. It would be good if we could introduce four or even five. That way, next spring, if one or more are not usable, hopefully two or three will still be good to go for the split.

Wax Moths

Linda asks:

I've heard that wax moths can be very destructive, but I have not seen what a wax moth looks like, nor have I seen the damage it can do. Can you show us an example?

Answer:

I was surprised and happy to see that we are not the only Beekeepers talking about wax moths before the coming harvest. You may find this forwarded email interesting.

From Bee Informed Partnership 11 Jul 2012 07:05 AM PDT

A female wax moth can produce up to 300 eggs and while not all will hatch, it is easy to see how a wax moth issue can quickly grow out of hand. Clothed in darkness, females typically lay their eggs at night, vanishing before daylight, but leaving evidence of destruction in their wake. After the eggs hatch and move into the larval stage, and in my opinion, grotesque, plump, soft bodied gray worms, about ¾ of an inch long, they burrow into the comb creating a series of tunnels underneath cappings and weave a series of sticky white webs. The larvae stay at this stage for about 3 weeks to 3 months depending on the temperature (the hotter the faster the infestation occurs) and then transition to their adult stage by way of a tough, silk cocoon.

During the daylight hours the moths generally hide in dark places carrying out their destructive behavior at night. The moths prefer comb that was once used for brood rearing, rarely damaging foundation but can be quite destructive to the wooden frame. A common misconception among beekeepers is that wax moths kill their colonies, but this is not so. A strong hive will have enough bees to fend off an infestation whereas a weak colony that is queenless, or has weak queen and a lesser population of bees cannot fend off wax moth larvae tunneling into and destroying comb. So while wax moth may contribute to the destruction of a colony, it is not the sole cause.

Here's something to remember about wax moths: They want brood comb. If you have healthy bees on your brood comb, then you are in good shape. Don't worry. If not, consider protecting your comb. Reduce the number of frames your weak colony

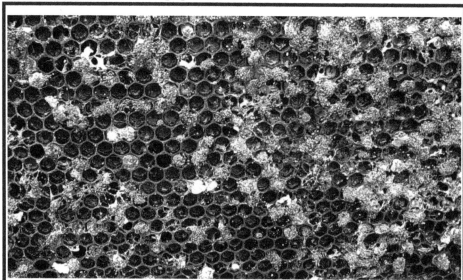

Illustration 144: Wax moth destruction. Photo credit: Don Studinski

must protect and move the rest to storage. You may have to move the colony to a single box or a nuc. I put my stored frames in plastic tubs. I get the clear plastic 56 Qt tubs. They fit 10 frames and you can see what is in there. Sweet.

Another option is to freeze your frames before you store them. An overnight freeze will kill all the wax moth eggs already there that you don't see. Otherwise, you may open up the tub in spring and find the wax moths have ruined your frames even though you put them in the tubs.

Purchased Bees

Bill writes about his bee investment this year:
Inspected the hives today; somewhat disappointing; particularly the vendor X bees.

Your swarm was by far the healthiest and have obviously been hard at work. They were busy on all seven frames, capped honey around the outside of the inner frames, capped brood was somewhat shotgun but larvae in the cells not capped.

The vendor X nuc was really disappointing, very little comb built, shotgun brood, very little capped honey, was somewhat concerned about pulling all the bars for inspection so left them alone.

Vendor Y package bees were also somewhat disappointing considering they have nearly a month over your swarm. Unlike your swarm they had not yet even touched the last couple of frames, shotgun brood, very little capped honey on the frames with brood.

Looks like swarms are the way to go, perhaps inexperience comes into play, but purchased bees seem to have a difficult time establishing themselves.

Answer:
These are not good brood patterns.

Some I can't see, but what I can see does not show good football shape nor wall to wall.

Illustration 145: Queen Street Bees. Photo credit: Bill Koeppen

Illustration 146: Vendor X nuc results. Photo credit: Bill Koeppen

If it was me, I'd be sending pics to Vendor X. The vendor needs to know how their colonies perform. That's <u>not</u> to say a refund is in order or even a complaint. But if I was the vendor, I would want to know so I might have an opportunity for improvement in the

Illustration 147: Vendor X nuc results: Photo credit: Bill Koeppen

future. These are live creatures. There can be no guarantees. This makes me wonder what happens in transit. I think it stresses the bees quite badly.

The swarm is also not showing good brood pattern. Still, it sounds like they are performing adequately given the difficult weather of late.

Also of note, Bill is experiencing the "after solstice" aggression. Take the time to light and use your smokers.

Pop Quiz

We performed an after-solstice split on Pumpkin 1 producing Pumpkin 1.1. Suppose we find a split failure. No brood, or perhaps a laying worker. Who can tell us what it looks like if we have a laying worker? Who can tell us what we should do if we see evidence of failure? Why can we get away with doing that? If we have split failure with Pumpkin 1, does that affect our plans for merge with Pumpkin 2 in August or September?

Answers:

Laying workers can only lay drones because they have no way to produce a fertile egg. Laying workers have inadequate abdomens for proper laying, therefore, eggs are not typically in the correct position in a cell and there may be double eggs in a cell. Many of their eggs will fail; this results in a widely scattered pattern.

If we see evidence of failure, we have these options: get a queen or merge with another colony. The obvious merge candidate is the mother colony. We can just put Pumpkin 1.1 right back on Pumpkin 1 because they are already sisters. No merge problems. Back to where they started.

Split failure has nothing to do with future merge plans other than possibly merging back to the mother colony.

Earwigs

Linda asks:

When I pulled the top off the quilt box, quite a few earwigs dropped out and were scampering for cover. I didn't see any earwigs in the hive, just in and around the quilt box. Is this a problem?

Answer: I see earwigs every year. Sometimes a lot. Sometimes a few. I am not aware of any reason to be concerned. I have never seen any damage caused. Some theorize that the earwigs may be eating mites, which would make them a good thing. Our permaculture knowledge tells us that there can be no doubt, they have found something in the hive that is worth while. That something might be the warm dry housing, or, it might be some food source. Although this is not clear to me, I am very clear on not worrying about them. They show up because nature says they need to be there. If they get too populous, nature will find a way to take them out.

Mid-Season Harvest

Rosemary writes:

I have a hive dilemma. My golden mean top-bar hive with the swarm is doing great. I actually need to go in ASAP because they have built comb all the way to the back.

Congratulations. Swarms are awesome. Consider this: although it is not "necessary" to harvest, it sounds like it would be fine to do so. Make sure it's still quite heavy to "heft" even after harvest to ensure they can make it through winter. I harvest TBH in spring rather than winter, sometimes, to ensure adequate winter stores. I suggest you don't take more than 30 lbs. You probably have about seven or eight pounds per frame.

My second package bee hive (again a golden mean top-bar hive) is doing okay. It is over 1/2 full of comb. I started feeding a little sugar water last week because they just started looking stressed- less active with the heat/drought.

I have become totally disenchanted with packages. We bring in disease and the bees don't perform. I will not buy any more. By feeding sugar water, you have contaminated any honey in there, not that it's bad, it's just not honey.

So my question is when I take the bars out of the swarm hive -assuming everything goes well and they don't fall off- could/should I put them in the package hive to give them a boost?
Could, yes. Should, I don't think so. Do you want to encourage substandard genes to occupy our precious Front Range? If they are going to survive, I think they should survive on their own.

If I did that is it difficult to get the bees from the swarm hive off to put the bars in the package hive?
If you move them into the package hive, then just brush off the bees as best you can. They are house keepers. If you put a few into the package, it's not a big deal. They might be accepted or they might be killed. One option is to spray them with Lavender water to disguise their smell.

Should I harvest the honey and save it to feed the colony if needed?
This is a perfectly logical option if you have a safe place for storage. This would give you the option to feed it to either colony. It may be smarter than feeding it to them now. They should be finding nectar now, especially in suburbia where people water as if water was cheap and easy.

I am planning on taking two bars out. The end comb is also turned so I may take it out as well?
Remove capped honey comb. Do not remove uncapped nectar, it will spoil.

The hive seems so full. Will they swarm midsummer?
Short answer: No.

More interesting answer:
The old saying goes like this
A swarm in May is worth a load of hay.
A swarm in June is worth a silver spoon.
A swarm in July ain't worth a fly!

Yes, they sometimes, rarely, swarm late in the season, but if they do that, it's pretty much suicide. Almost certain suicide for the swarm, they simply do not have enough time to build the necessary stores to get through winter. Could also be suicide for the mother colony; greatly diminished population available for finishing winter preparation plus they have to wait three weeks for the new queen to begin laying. Let's run through the math again:
old queen leaves upon new queen being capped, that's day 9
new queen emerges on day 16, one week later
new queen must sexually mature ... another week, that's 2
new queen must fly and mate with 10-20 drones ... another week, that's 3
now new queen can begin laying.
New brood will not begin emerging for yet another 21 days!
It's a huge set back.

A summer worker bee lives for 6 weeks. Think about that. What does this do to the worker population?

Plus, I ask you, if they swarm in July, August, Sept, I've even heard of it in October, but I find that very hard to believe, is that really genes we want? Learn to let them go. We can't save them all. It sounds cruel, but it's not, it's just nature. And in the end, it's the kindest way! Think about how many fewer bees will be suffering from inadequate characteristics if we let them go as soon as nature says so. I love the book, Ishmael! If you have not read it, you owe it to yourself to read it. Anyway, as Ishmael would say: "These are god decisions. Leave them to the gods."

Wisconsin Questions

Did They Swarm?

Robin writes:

I do have a question for you. I'm pretty sure both my hives swarmed. I didn't have enough boxes (didn't realize how fast they would need them) and had a delay in giving them more room.

Two deeps should be enough room. This is late for swarming. It's also unusual for a newly installed nuc to swarm in the first year. Maybe you are having an outstanding flow and they expanded dramatically.

That's what I initially thought too, that two deeps would be enough until next spring and then I was going to split them. That's why the delay with having more boxes. They did seem to expand dramatically, but I just thought they always do that and I just wasn't experienced enough to know it. I know our climates are entirely different, but do you think by swarming this late they will be less likely to survive the winter?

Yes. I guess it would have been good to load them up with supers. You could have had a nice harvest.

They may be queenless. I'm worried. I can't find the queen.

Do you have capped brood? Is it all drone? Does it look like a laying worker pattern?

So basically what you mean here is that I am panicking too soon. There was definitely capped brood in a very nice pattern with pollen (is that called bee bread when it's capped?) and honey over top.

Yes, I think you are too worried. Yes, pollen is called bee bread after enzymes and honey have been added and it is packed into cells.

We were so worried that we would hurt the queen cells and end up with no queens that we didn't dig too deep into the hives.

When you are trying for queen cells, yes, it is very important to move horizontally before vertically on any inside frame, outside frames can move straight up. However, when you are trying to avoid a swarm, killing queen cells is not a problem. Still, my advice is to avoid swarms <u>way</u> in advance.

We didn't see any queen cells in the other one, but we were only into the top box, we didn't check the bottom one (they were two box hives, we added a box to the more populous hive and will be adding one to the other one this week).

I run three deeps on many of my Langs. I'm trying for a colony of 100,000 bees. Are you trying for three deeps intentionally?

I don't know if I'm really trying for 3 deeps, I want to do an unlimited brood nest management system and the bees seemed to need the room. There does seem to be a lot of honey in there, but I don't know if they are honey bound. That means that the brood nest is stuffed with honey also, right? How many frames will they use for brood in a 10 frame Lang? If you don't mind, maybe you could explain the concept of an unlimited brood nest, I think I'm trying to do something I don't fully understand.

I'll discuss unlimited brood in another lesson about three deep management. Yes, honey bound means they have put up so much honey they are running out of room for brood and perhaps have used some of the brood area for storing honey. There is nowhere left to go with brood, thus the term "bound" as in tied up. This can indeed cause them to swarm. Still, this is unusual this time of year. The number of frames of brood in any hive depends on the queen's ability to lay and the hive mind's directions to expand, maintain or contract the population. They could be using all 10 frames in a deep or much less. Did you look at the picture of queen a cell? Is that what you saw? Were they empty?

Do you mean to load a super on top of the 3 deeps?

I do this, in fact, I currently have 4 supers on my strongest three deep hive. I am not, necessarily, recommending this for your current situation. If you think the flow is strong enough and you want honey to harvest, put a super on the 2 deeps and see what you get. No queen excluder is necessary. Otherwise, you can just put another deep on ... now you are at 3 deeps. See what they build. Bottom deep will be empty (of bees) come spring.

And if they fill the super, then harvest it at the first spring inspection?
No. if they fill the super, harvest it this fall. That should still leave them 60-90 lbs of honey and bee bread in the 2 deeps. You need to confirm this the 60-90 lbs in the fall.

If that is what you mean, would you recommend I use a queen excluder, or is it unlikely she will go up that far?
Queens will not traverse solid spans of honey to lay. The hive mind wants the brood together both vertically and horizontally. No need to exclude. They already have a solid span of honey across the top of your upper deep.

Some of the frames were stuck to the frames in the bottom box. That's why we were so worried about hurting the queen cells.
This time of year, I would expect to see <u>drone</u> comb in this position. Yes, it could be queen cells for swarming, but why are they swarming now ... it's almost certainly suicide. Are they honey bound?

I do think a lot of it was drone comb, are the queen cells dependably on the actual frames and not in burr comb? I wasn't sure when we were out there and since it seemed they had swarmed with no young larvae or eggs present and therefore needed to make another queen ASAP I didn't want to risk their only chance. Do you think its suicidal for the ones left, the ones that swarmed, or both?
Queen cells can be found anywhere. It depends on hive mind motivation. If the motivation is swarm, then they have plenty of time to plan it out and they will build the cell toward the bottom of the frame and tell the existing queen when to fill it with an egg. If the motivation is supersedure, then the existing queen is failing. They may need to use a viable egg that has already been laid in a mid-frame position or really <u>anywhere</u> on the frame. What makes you think they have swarmed? I'm thinking they have not. Swarming this time of year is almost certainly suicide for the swarm. They do not have enough time to find a new home and load it with stores for winter. It could be suicide for the mother colony too if they lose so much population that they cannot adequately defend their stores (they could get robbed to death) or if they cannot cluster adequately come winter. If you still have a hive full of bees, 10 or 15 frames out of 20, then they have not swarmed.

The bottom boxes on each hive both still have the nuc frames in them. The people I bought the nucs from had them very spread out so the bees built them really thick. I could only fit the five nuc frames and four foundation frames in there when we set them up. When we looked down in between the frames we pulled out, it looks like we are going to have to cut the bottom box frames apart to get them out. They seem to be built together.
Getting those thick frames out is a brutal manipulation. There is no way around it. When the time is right, you just have to slice comb with your hive tool and get the frames out. You may have to scrape a lot of comb off the frames. Bring a tub for honey and a separate tub for killed brood. Chickens will love to eat the brood. You can melt the wax to make soap, etc. I ran into this with Pumpkin 1 and Pumpkin 2 also. Some beeks just don't bother learning or they think they know so much better than the 100s of years of experience brought by Langstroth and his followers. There is room for 10 frames in there for a reason. On my Pumpkin frames I ended up trashing a lot of honey and brood to get that thick comb out of there. I did that early in the year so the bees could recover. You will have to assess whether you need

to do it now, or wait until next year.

So I should leave them alone long enough that a new queen will have time to emerge, mature, mate and start laying. I have a magnifying glass to help me look for eggs, we didn't find any. Then once laying is verified I can start my "brutal manipulation" of the bottom boxes. Right? Can you give me a rough idea of the time I need to wait before knowing for sure if they've requeened or not?

If they have indeed swarmed, which I doubt, then you can confirm this, the timing is very tight. The last eggs laid by the old queen will emerge 21 days after she left. We want to look for "no capped brood." New queen was capped when old queen left. She will emerge 7 days after capped. She must mature about a week to be ready sexually, that's 14 days. She must fly and mate with 10-20 drones which takes about a week, that's 21 days. Then, she starts laying. This means the new queen begins to lay about the same time the last of the old queens daughters are emerging. But, those new eggs will not be capped for 9 days. That gives you about 7 to 9 days to check for "no capped brood." If you see that condition in those days then they did swarm and you have a new queen. Or, you might be queenless. If you don't see that condition in that window of time, then they did not swarm. Old queen is still there and laying.

Oh, one more thing. Hopefully it won't be a catastrophic mistake, but when I put the 3rd deep on the one hive, I brought up every other frame in the second deep to the third (in the same position) and filled in what I took out of the second and the rest of the third with foundationless frames. After reading what you've been teaching me, it sounds like that would be splitting the brood nest up too much. Please advise on how to add the 3rd deep to the other hive more correctly. Is there anything I should do to correct what I already did?

Yes, it's a mistake. This is very disruptive to their brood nest. You split it across 2 levels vertically and across empty frames horizontally. You probably gave them 8 different locations to keep brood warm. At this point, it's too late, so just leave it. Here's why. One of 2 scenarios played out.

1. They couldn't keep all the brood temperature controlled and some is now dead. They will abandon it. Eventually, they will clear out the bodies and reuse the drawn comb in the most efficient manner.

2. The brood temperature worked out and it lived. Those bees will emerge and help to build new comb on the new frames. This is likely your case because you say the hive is so full of bees.

Leave it. Try to disturb them as little as possible.

North Carolina Questions

Newbie Perform Removal?

Linda writes:

I have a colony I need to remove and I'm trying to get ideas on how. Here is the situation. They are inside a church fellowship hall, they are going and coming under the siding on the back side of the building. The building had a flat roof, but that roof is now pitched. I can go inside the attic, at that time I'm walking on the old roof under the new one. From what I can see the nest is between the old roof and the ceiling. They did not want to mess with them but they tell me there is black stuff on the ceiling and wall where the bees are. From what I gather they have been there for several years.

It's expensive. I would have to fly there with my tools and perform a removal which is not trivial. There will be wounds to the building which must be repaired by someone else.

I have a fellow bee keeper willing to help me perform this, we are just looking for ideas of how to get it done.

I'm sorry, but trying to coach someone through a removal remotely is beyond the intent of the mentor program. I'm not even sure it could be done. New beekeepers should not be attempting structural removal. You need expertise with the bees, expertise with construction and expertise with a wide range of tools.

I was doing an inspection on the queen Kierra hive and found in the brood box I have two frames that have come apart. How do I fix that?

Eventually the time will come to cycle out those broken frames. At that point you can repair them or toss them to the land fill.

No Poison Required

© steve kennedy 2014

In *Bee Culture*, January, 2013, a very experienced beekeeper, Ed, laments the fact that 61% of beekeepers are using <u>no chemicals</u> to treat varroa mites. To Ed, that seems an illogical choice. To me, it seems not only logical, but necessary. Let me explain.

Illustration 148: Varroa destructor on honeybee host. Photo credit: public domain, USDA ARS, from wikimedia

The EPA defines *pests* to include some plants, some insects and some mammals. That is to say, when you are applying an herbicide, like Roundup®, you are using a pesticide. Here, we will discuss pesticides in general (poisons that kill plants, insects and mammals), even though I may refer to specific examples along the way. The *pesticide spiral* is a concept that applies to <u>any</u> pesticide without exception. It's an acknowledgment of how nature and genetics react to poisons in general. Setting the stage, we should note that, according to the EPA, about 5.1 billion pounds of pesticides are used, just in the United States, each year.[40] About 95% of soybean and cotton crops and more than 85% of corn in the U.S. are planted in varieties genetically modified to be herbicide resistant.

Recent findings show that Roundup®, also known as glyphosate, in combination with Roundup® ready crops, like corn, soy and cotton, is a strategy that has failed. Rather than these handy tools reducing the amount of poison we are spreading over the Earth's surface (as originally promoted and sold), we are

40 http://www.epa.gov/pesticides/factsheets/securty.htm, visited 8/14/2014

now using more pesticides than ever.[41] This is certainly good news for the companies that sell us the poisons.

What is nature's reaction to poison? Let's see how the cycle works. It's fully understood by the poison manufacturers, but obviously, they do not want this knowledge widely promoted.

Our media outlets leave us with the impression that using things like antibiotics actually "causes" antibiotic resistant germs (pests) to develop. This is not true. Using a poison does not induce the evolution of a species. Evolution is always underway without regard to what poisons are being used or not used at any given time. Nature produces a variety of unique species modifications. The survivors carry their genes forward to future generations; others do not. The evolved antibiotic resistant germs already existed. It's just the population that changes.

Consider a fictional example which will make the concept easy to understand.

Illustration 149: Flying dog. Photo credit: public domain, from http://northbayanimalhospital.blogspot.com

Let's use those darn, pesky flying dogs to illustrate. Someone develops a poison that kills flying dogs. We start spreading this poison over an area, for example, Colorado. This poison kills all flying dogs in Colorado that are vulnerable to that specific poison. What does that leave behind? Answer: all flying dogs that are not vulnerable to that specific poison. "Duh!" you say. Me too. This leaves behind two important characteristics. First, a set of flying dogs with genes that can live in the presence of the poison. Second, the entire food supply for the flying dogs that remain. All their competitor flying dogs have been killed. The food supply which had previously been shared by many diverse genetic combinations is now available to just the few genetic combinations that were not vulnerable to the poison. What always happens when there is an abundance of food? Answer: the population of creatures that feed on that source expands to take advantage of the food. In plenty, more offspring live. In this case, offspring of flying dogs that can successfully live in the presence of the poison.

Bear with me, I will get to the connection to honeybees.

This leaves us with a large increase, in Colorado, of pesticide resistant flying dogs. It didn't cause the flying dogs to become resistant. It simply increased the population of a specific set of genes by giving them the entire food supply. By the way, assuming we didn't apply this poison in Kansas, the genes over in Kansas continued to evolve as before. The flying dogs that are resistant to this pesticide are there, but in much smaller numbers. They are sharing the food supply with all of nature's resilient diversity of gene combinations.

Now let's look at the Roundup® case. Supposedly, the use of Roundup® "caused" Roundup®-resistant weeds to develop. No, that's not correct. The use of Roundup® created an environment where the Roundup®-resistant weeds could thrive: no competition for nutrients. So, where historically, US farmers were using 1.5 million pounds (1999) of herbicides to kill Roundup-resistant weeds, now they

---

41 http://gmwatch.org/latest-listing/51-2012/14041-new-benbrook-data-blow-away-claims-of-pesticide-reduction-due-to-gm-crops, visited 8/14/2014

are using 90 million pounds (2011).[42] Note that we didn't say any less Roundup® was being used. We just said that the herbicides necessary to kill Roundup®-resistant plants increased by 60X in 12 years. Think the chemical companies are pretty happy about that? I do.

We all need to be applying this knowledge to our interaction with honeybees. If we treat for mites using poison, are we reducing the mite problem? No, we are actually shifting our problem from those mites that are <u>vulnerable</u> to poison to those mites that are <u>resistant</u> to treatment. In the short run, you may reduce your mite population. But that reduced population consists of those mites resistant to treatment and they now have the entire food supply. In the long run, you are subjecting your colonies to the survivors; the most poison resistant mites. In either case, <u>honeybees that can cope with mites will ultimately survive</u>. The question becomes: how long will it take us to get to where our honeybee population is <u>dominated</u> by those that are successful living with mites? The longer we spend "defending" our honeybees from mites, the longer we drag out eliminating the honeybee genes that cannot cope with mites. As those that cannot cope with mites are eliminated, it leaves a larger food supply available to those that can cope with mites, and consequently, their population will expand.

Illustration 150: Apistan strips (varroa mite poison) being installed. Look closely, they are vertical clear strips of plastic about one inch wide. Photo credit: Dave Cushman website

If we want to have honeybees that can live successfully in the presence of mites, the most likely successful and healthy path is to allow nature to choose them. In other words, we will have to experience the loss of colonies not able to coexist with the mites.

If we use poison, miticide, organic or synthetic (it doesn't matter), to control the mites, then we are stepping into the pesticide spiral. We are selecting for survivors in the presence of _____; you fill in the blank. This same law of nature holds true for any honeybee disease, parasite or competitor. When we chemically manipulate the environment to change the selected survivors, we will, eventually, need a different (more powerful) chemical to continue that manipulation. That's the pesticide spiral. The only way out is to let nature run its course and naturally select for balance (diversity) of individuals within species.

Nature is a highly complex system that will, ultimately, be successful in selecting survivor stock. Take a look at this example. It's so beautiful how nature can invent these incredible relationships.

> *Quercus lobata* Née, the Valley Oak tree, is a late summer source of bee forage in Northern California where forage is scarce going into the fall. Valley oaks are endemic to California and are found in the interior valleys and foothills. At this time of year one can hear honey bees buzzing high up in the canopy. They're not visiting flowers, but "oak apples," a type of gall induced by the oak gall wasp *Andricus quercuscalifornicus*. What the bees are after is the

Illustration 151: Andricus quercuscalifornicus, Valley Oak Gall. Photo Credit: public domain, Hank Fabian, http://bayareabiologycourses.weebly.com

honeydew the gall secretes.

The tiny wasp responsible for inducing these galls lays her eggs in plant tissue on the stems of Valley Oaks. The emerging larvae stimulates the growth of a gall which acts as a nutrient sink providing the growing wasp with the food it needs throughout its development. As a result of nectar secretion, the gall attracts bees, ants, flies and beetles. When these insects feed on the honeydew they discourage parasitoid wasps from disturbing the developing *A.quercuscalifornicus* wasp within the gall.[43]

I frequently find myself saying to students phrases like "we don't know" or "it's not knowable." When I say these things, it's the symbiotic relationships like the gall and the wasp on my mind. When we do something like introduce systemic pesticides to our environment, the most widely used poisons in the world today, we don't know the full effect because it's not knowable.

The wise beekeeper allows nature to work its unknown and unknowable magic. Yes, you will lose colonies of bees. Be of good cheer. You are one step closer to a larger population of honeybees that will be successful. When we do something like introduce miticides to our environment, we don't know the full effect because it's not knowable. Which honeybees shall we allow to survive? This decision is best left to the gods.

Here's some good news. In the December, 2012 issue of *Bee Culture*, page 81, Alan Harman reports that the honeybee "genes associated with worker behavior were found in areas of the genome that have the highest rate of recombination." Recombination is the shuffling of the genetic deck occurring in the queen's ovaries as eggs are created. Alan further quotes Proceedings of the National Academy of Sciences, honeybees have "the highest rates of recombination in animals - 10 times higher than humans." This increases the chances that honeybees will hit on the right combination of genes to successfully co-exist with mites. More recombination means more diversity. From that diversity, natural selection can determine what works. More combinations gives more chances for success.

Some will argue: Agriculture cannot be successful on the scale necessary today without killing pests. This is simply not true. Consider the effect of these small and natural changes to production scale agriculture.

Longer rotations produced better yields of both corn and soy, reduced the need for nitrogen fertilizer and herbicides by up to 88 percent, reduced the amounts of toxins in groundwater 200-fold and didn't reduce profits by a single cent.[44] Adam Davis, an author of the study who works for the U.S.D.A., said,

---

43 Correspondence from Bee.Health.eXtension@gmail.com 9/18/2012 @ 1614
44 http://opinionator.blogs.nytimes.com/2012/10/19/a-simple-fix-for-food/?_php=true&_type=blogs&_r=0, visited 8/14/2014

"These were simple changes patterned after those used by North American farmers for generations. What we found was that if you don't hold the natural forces back they are going to work for you."

So, it turns out, the 61% of beekeepers that are not using miticides to kill varroa destructor are actually doing us all a favor. Life gets easier when we start trusting nature and working with it instead of fighting it with chemicals to fit some invented notion of "what should be."

August ... Harvest Preparation

© steve kennedy 2014

Harvest is approaching. The classic harvest timing for Colorado's Front Range is end of August or early September. There certainly can be exceptions that make a lot of sense. For example, many beekeepers that manage top-bar or Warré hives may choose a spring harvest. The logic is to leave the stores for the cold weather season to ensure the bees have adequate food. If there is honey left in spring it can be taken because the abundant nectar flow is happening or will soon happen and the bees can easily restock. Another legitimate choice is to harvest at summer solstice. In this case, the idea is that the major honey flow is done and it's more convenient to harvest in warm weather while the honey flows easily out of the comb. This would allow the bees plenty of time to build up for winter after the beekeeper has taken their share. But the typical choice is right around fall equinox. This allows the beekeeper to ensure that adequate cold season stores are in place before they take their harvest. This also lines up with the many harvest festivals that communities enjoy and allows the beekeeper to quickly sell their fresh honey.

This can be a substantial intrusion into the hive. It's the time of year when the bees are comparatively more defensive and smoke is a must. Bees are not programmed to be happy about you taking their stores even when you are leaving plenty behind. That's their hard work you are taking away and they know it can mean the difference between life and death for them in the coming months. You should actually be glad they are defensive because it means they are doing their job, just as you would want them to do if you were not around.

What You Need

While the weather is warm and pleasant, pick a place where you will store your used, empty supers during the cold season. That place should be thoroughly cleaned and ready to receive those supers when the day comes. The place should be safe from pests that would like honey residue or wax. You will be glad you prepared this place in advance. Harvest is busy. You will not want to take a break to do this in the middle of harvest.

Illustration 152: Freshly washed buckets drying in the sun. Photo credit: Don Studinski

For harvest, you will need:

1. Frame sized plastic tubs to hold harvest frames or large, tough garbage bags to hold supers.
2. A wax cutting knife for cutting comb off top-bar and Warré-style frames.
3. A wax cap removing tool. This might be an official wax cap removing heated knife, a capping scraper or it might be just a bread knife.
4. Food grade five-gallon buckets, one for every 50 pounds of harvest, plus an extra just to be safe.
5. At least one food grade **gated** five-gallon bucket.

6. A strainer and a strategy for cleaning it out when it gets clogged. It helps to have two.
7. Rags wet and dry.
8. A good spatula.
9. A butter knife to catch drips.
10. An extractor for spinning frames or a crushed comb setup for using gravity.
11. Jars to receive the end result. I prefer a wide mouthed jar for when the honey crystallizes which almost always happens to raw honey eventually.

Some Front Range beekeepers have supers full of honey. Others do not. I hear of decent harvest for beekeepers in Hygiene, but not much coming out of Aurora. Your harvest will vary based on your **micro climate**.

A beekeeper saying goes like this, you can make bees or you can make honey, but you can't make both. Some years I concentrate on multiplying my living colonies without expectation of honey harvest. This year I was expecting harvest from a few dedicated colonies. For any given colony, you must make this decision early in the year and live with it. You may be able to make honey with some colonies and bees with others. As it turns out, the year has presented us with some pretty significant weather challenges with extreme heat and lack of water. The hives I dedicated to building a honey harvest have produced, but much smaller than expected. Oh well, this is agriculture; there are no guarantees.

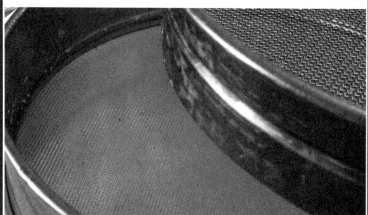

Illustration 153: Stainless double strainer. Photo credit: Don Studinski

If you have supers on, you should be preparing to remove them.

How can you know how much honey you can take as your harvest?

In our climate, honeybees need between 60 and 90 pounds of honey to get through winter. My approach to testing for adequate stores is to perform what is called the "heft test." This means I approach the hive from the rear and lift the back slightly, leaving the front sitting on the stand. You will develop a "feel" for a hive ready for winter. A good way to practice this is to go to the dog food aisle at the grocery store and practice lifting 50 pound bags of dog food several times until your arms get the feel for that amount of weight. Then go to your hive and do the heft test. If it feels about the same, you are good to go.

If, after you have removed your supers, your hive feels light, then some of that honey should be set aside. Your bees can still gather nectar and produce honey through the end of October. Should your hive fail to increase in weight adequately for winter, you can put some of that honey back. Remember, honey should be on the outside of the brood nest. Do not break the brood nest into two separate areas in October. Further, do not leave a queen excluder on over winter, even if you leave on a super to feed the bees.

Having pulled the supers off and gotten some idea about how much honey you will actually take as your reward, it's time to get ready for a big sticky chore.

You will need containers (perhaps the supers themselves) to hold your harvest frames. They need to be protected from pests while they are awaiting extraction. Some use a plastic garbage bag over the super.

I use plastic tubs with lids that will fit the frames. The trick is getting the bees off the frames before you store them. It is not realistic to think you will get every bee off those frames.

You will need a means of removing wax caps. I have a plastic tub with a wooden bar that reaches side to side. I can rest the frame on the wooden bar and cut the caps off. They fall into the tub. Later, I can melt the caps and separate the residue honey from the caps. That is generally a chore for a different day.

You will need a way of extracting the honey from the frame. The quickest way is a fancy electric extractor that spins the frame until nearly all the honey has flown out and drips out the bottom of the extractor. Crushed comb is another alternative. In the crushed comb case, you crush the comb so the honey can drip out through a strainer. The strainer captures the comb and bee bits and the honey is captured clean in the bottom. I also use a strainer for the honey coming out of the extractor. In every case, I highly recommend you think of a way to be able to temporarily remove the strainer for a quick clean out of the debris because it will get clogged up as you progress.

You will need to have five-gallon food grade buckets to receive your honey. Each one can hold about 50 pounds of honey depending on how close to the top you can fill it (the strainer tends to prevent getting it full to the top because it hangs down into the bucket). At least one of your buckets needs to have a "gate" which is the spout where you can let out honey one jar at a time.

You will need to have purchased jars and lids for your honey. Loading the jars can be a different day from extracting and probably should be.

Illustration 154: Gated five-gallon bucket. Photo credit: Don Studinski

All this stuff has to be washed and prepared for food in advance of your chores. Then, it has to remain clean until you actually get to using it.

Example Schedule

Here's the sequence of steps:

1. Clean your extraction area and get it organized.
2. Prepare your extraction containers and tools.
3. Remove supers and queen excluder.
4. Test hive weight.
5. Set aside honey to feed back to the bees.
6. Safely store honey to be extracted.
7. Perform honey extraction into five-gallon buckets.
8. Potentially return supers to hive to let the bees clean them up.
9. Fill honey jars from five-gallon buckets.
10. Remove the supers again and store them for the off season.

It's a good idea to harvest early enough that you can put the wet frames back on the hive to let the bees clean them up before it gets too cold. They will need a week or two for this clean up work. Then you will want to store those supers in a place where they are protected from pests so you will be able to reuse them next year.

Here's an example schedule that you might find useful.

Saturday, 9/1

Starting at Hearteye Village, check Clarkson, TBH, expect 3 frames in 1 tub
Move to Rock Creek, check Pumpkin 1 and 2, expect 2 supers in 2 tubs
Leave empty super boxes on site.
Move to Living Systems Institute, store tubs, thorough clean of garage, set up crush press, it must be sitting high enough to drain into five-gallon bucket.

Illustration 155: Honey harvest in a frame sized tub. Photo credit: Don Studinski

Sunday, 9/2

Starting at Hygiene, check TBH, expect 3 frames in 1 tub
Move to Boulder, Mica, check TBH, expect 3 frames in same tub
Move to Boulder, Leslie, check Quay, Warré, expect 1 super in 1 tub
Leave empty super box on site.
Move to Living Systems Institute, store tubs
Move to Denver Urban Homesteading, finalize spin setup

Monday, 9/3, half day ... holiday

Starting at Crescent Grange, check Sugarloaf, Birch 2 & Birch 3, expect 3 supers in 3 tubs
Leave empty super boxes on site.
Move to Living Systems Institute, store tubs

Tuesday, 9/4

Starting at Golden, Rinehart, check Wiley, Warré, expect 1 super in 1 tub
Check Rinehart, expect 1 super in 1 tub
Leave empty super boxes on site.
Move to Living Systems Institute, store tubs
Move to Lakewood, Whitehead, check Newport, Warré, expect 1 super in same tub
Check QE II, expect 1 super in 1 tub
Leave empty super boxes on site.
Move to Living Systems Institute, store tubs

Wednesday, 9/5

Starting at Living Systems Institute, check Sable, TBH, expect 3 frames in existing tub

Check Superior, expect 2 supers in 2 tubs
Check Braden, expect 1 super in 1 tub
Check Birch, expect 4 supers in 4 tubs

Four tubs contain comb that can be crushed. Free these up using six five-gallon buckets. Reuse tubs for Lang frames.

Thirteen tubs contain Lang frames.

Must crush before we spin in order to free up 5 gal buckets to catch spun honey.

Thursday, 9/6

Crush day at Living Systems Institute.
Six buckets of crushable comb.
Some may be saved as eatable comb honey … only that which is very pretty.

Friday, 9/7

Move all 13 tubs to Denver Urban Homesteading.
Spin honey out of all frames.
Six to ten 5-gal buckets of honey is yield.

Saturday, 9/8

Bottle honey at Living Systems Institute.
Need about 300 bottles.

Bizarre Weather Impacts Bees

Beekeeping is living with nature much the same as ranching. In fact, bees are considered livestock, same as cows.

Sometimes, things are good and harvest is plentiful.

Sometimes, things are bad and you lose your stock.

Right now, 8/1/2012, many ranchers are selling their cows earlier than expected due to drought. Sell now at a loss or take the chance on losing it all to death.

Likewise, some beekeepers are seeing poor results at this time. The exact cause may or may not be known. But one thing is clear: Poor beekeeping results does not necessarily imply poor beekeeping.

There are many reasons to get discouraged as a beekeeper, but experiencing a failing colony is not one we should allow. This is

Illustration 156: Washing all the tubs. Photo credit: Don Studinski

going to happen. We must be prepared to accept it, learn what we can, and move on.

Here's some stuff you might appreciate if you are feeling discouraged:

Most of Bill's package bees are dead. They were doing great, but suddenly they faded away. We merged them with the Queen Street removal bees, which was also doing great. Now they are building queen cells. Why? They are telling us that their queen is failing.

Wynn grabbed a healthy looking swarm. They showed up queenless a couple weeks ago and had to be merged unexpectedly.

Susan got a weak swarm. They turned out to be queenless. Susan invested in a queen and introduced it over four days. They are still failing.

Leslie has been successful with the queen cell we provided. Laying is strong. Still, that colony is building queen cells.

Bobbi has struggled with a removal colony. First cycle left the queen behind. Second cycle merged this queenless bunch with the Maylan Street swarm which was small but healthy. Third cycle found the original queen and merged that. Now they appear to be queenless.

Illustration 157: Forgotten and on their own. Photo Credit: Mo McKenna

Robin is struggling with bees that make a mess of the frames and attach comb all over the place.

These are all normal abnormalities. You are not alone. You are not failing. You are learning.

Relax, the bees are rooting for us.

How Much Intervention?

Mo found this abandoned hive literally boiling over with bees on a farm she works. Though she has bees there, she had not noticed this particular hive which was a bit off the beaten path. She stuck a deep body with 10 clean frames on top of them.

It's crazy that we are all obsessing over our bees and these ladies are doing great, forgotten and on their own.

This is evidence that we don't know everything about keeping bees alive. Maybe, just maybe, human intervention is not necessary; perhaps even counter productive. Another beekeeper saying is: if you are not sure what to do, doing nothing is probably the best choice.

Getting Bees Off Your Harvest Frames

One method beekeepers have used to get the bees out of their harvest supers is called a **fume board**. It's a cover with a felt lining which can be used to hold liquid fragrance that chases the bees out of the supers and down into the brood area. It's chemicals. The last thing I would want to do is create my honey in a chemical free environment and then use chemicals to harvest it.

Another option is an **escape board**. It's an inner cover that allows bees to exit the supers down into the brood area, but not return. It takes quite some time to get your supers empty of bees, but it can work.

My preferred method is this: we smoke the bees to keep them calm. Then we pull one frame at a time and shake the bees into the hive. Some bees may need to be brushed off with your **bee brush**. Put the frame into a tub and close the lid. Very few bees make it into the tub. Move to next frame. This doesn't take long. Close up the hive. Haul away the tub.

Three Deeps - Why and How

One of my main objectives is to get each Langstroth colony to three deeps for brood. A colony this size is hitting about 100,000 honeybees in June or July. Daylight traffic at the front door should be dozens of bees each second. A huge number of foraging workers.

This is a foraging powerhouse and, therefore, a pollination dream. Almond farmers insist upon eight frames of bees per colony when they are paying rental fees for pollination service. The colony I'm describing will have 20 or more frames of bees. On a good flow, a three deep colony that is fully utilizing their space can fill four or even six supers with honey. Being robbed is out of the question.

Of course, finding the queen is also probably out of the question. A superorganism like this will tend to be healthy. Should individual members become injured or

Illustration 158: Superior colony. Three deeps plus supers. Photo credit: David Braden

sick, they can be eliminated without a noticeable affect on the whole.

I want colonies like this because they can bring in a harvest; plenty of food for the bees, both pollen and honey, and surplus for me. I also want this configuration because it aids in my comb rotation.

We know that some chemicals accumulate in the comb. Some well seasoned beekeepers recommend that we rotate out any comb still in use after three full seasons. It's inconvenient to have to sacrifice brood or food in order to cycle out comb. This inconvenient sacrifice can be avoided, but only by paying attention to the idea of comb rotation all year.

I find the most convenient time and place to cycle out comb is first inspection, removing the bottom of three deeps. Because I'm using the three deep configuration, I am confident that the bottom deep will be empty come first inspection, in March. The bees want their winter stores above their brood. That's why we are able to get surplus in our supers, far over their brood area. As the winter progresses, the honeybees will eat their way up into their stores in the upper two deeps. They may have not eaten all the way into the top deep, thus leaving the middle deep either occupied or empty. However, they will always have eaten far enough to leave the bottom deep empty. No brood and no food will be left. This is the time to cycle out that comb. You can remove the entire deep and replace it, on top, with new frames ready to be drawn out. The (formerly) middle deep becomes the new bottom deep. The (formerly) top deep becomes the new middle deep. The frames being cycled out can be moved to your **solar melter**. After they are cleaned up, they can be reused. If you repeat this cycle every year, then your frames are used for three years and cycled out on the fourth year.

What about all those Warré hives? Aren't we supposed to introduce a new deep at the bottom of those? Well, yes. The three deep discussion is a bit different for a Warré, but the motivations are the same: plenty of room for brood, target 100,000 bees and rotate out old comb.

Warré hives have a natural comb rotation built in the suggested management. There really is nothing equivalent to a super. All the boxes are deeps. In the spring, place an empty deep on the bottom board. All other deeps are stacked on top in the same order you found them. At harvest, pull your honey harvest off the top, but make sure you leave enough honey to get the ladies through winter. Harvested honey will be **cut comb**, therefore, the comb will not be used more than three seasons before it gets cut off.

Then there's the top-bar case. Rotate comb out of a top-bar hive in a manner similar to a Warré hive. This will happen at harvest time. Find their brood area. It will be either nearest the sunlight, or nearest the front door. Pull the top bars farthest away from that area. Those will be full of only honey. Cut that comb off the top bar for your harvest. Cycle that top bar into the brood area to get the bees to build new comb. In this manner, comb can be slowly shifted out to the harvest position such that some will be cycled out each year. Don't split up the brood area with more than one empty bar at a time. Placing an empty bar between two with straight drawn comb will force the bees to draw the empty one straight also.

Illustration 159: Mouse guard for top-bar hive. Photo credit: Don Studinski

Mouse Guard

During first inspection in 2012, we got to see first-hand the devastation that a mouse can perform on a honeybee colony. Mice find a heated bee hive (heated by bees) a very inviting place to camp out in winter, especially considering that a nice food source is also included.

The top-bar hive out at Broomfield Organic Garden contained a colony I had worried about all winter. I was watching the entrance on warm winter days. Sometimes one or two bees. Sometimes none. I had just about given up on them when, suddenly, their population burst forth. She was laying heavy. They had made it through the winter of 2011 / 2012! Unfortunately, the celebration would not last.

When we showed up for first inspection, we found a big tunnel of destroyed comb across the bottom of the hive. That tunnel was 4 inches diameter at the biggest spot, missing comb. The mouse ate bees, comb, brood, honey and bee bread. It apparently either ate or killed the queen. It left urine and feces everywhere. Within 2 weeks, the colony was dead. Lost food supply and small population were such powerful disadvantages, they could not survive the late winter temperatures of March.

That hive had only one possible entrance for a mouse. Three holes drilled into the side of the hive, each 3/4 inch diameter. These had all been somewhat reduced with burr comb. Think a mouse cannot squeeze through that 3/8 inch entrance in front? Think again. Mice can flatten themselves out and squeeze through an amazingly small space. Luckily for us, they cannot (read "I have not seen it") make it through a space of 1/2 inch by 1/2 inch square. That is the dimension of the hardware cloth I'm using for my **mouse guard**.

On the top-bar hives, the hardware cloth is placed directly over the entrance holes. On the Warré hives, we integrate the guard into the bottom board. For the Langstroth hives, we have rigged a small strip of hardware cloth attached to two small blocks of wood. The whole tool is exactly the width of the entrance. The two small blocks of wood are held in place by the bottom deep and the blocks of wood hold the hardware cloth in place across the entrance. This hardware cloth has been cut down to only two squares deep to cover the front entrance. One example has blocks 1/2" while the other has blocks 3/8". This is for the two types of bottom boards you will find in the Langstroth world.

Illustration 160: Mouse guard on Warré. Photo credit: Don Studinski

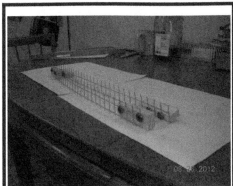

Illustration 161: Mouse guards for Langstroth hives. Photo caption: Don Studinski

I try to disturb my bees as little as possible. Combining tasks helps. So, I want to introduce the mouse guard while I am there doing something else. In this case, that something else is harvesting. Next week is harvest. I will have made all my mouse guards to the correct dimensions (for each hive) before I start. I will have them with me when I arrive. And I will get them in place before I leave.

You should decide how you will prevent mice from entering your hive. A friend told me, yesterday, they are already seeing mice in their house. Figure it out now. Keep the mouse guard on all fall and all winter at a minimum, maybe all year long. Mice are actively seeking a warm place to live in fall and winter. By spring, they may be ready to move back outside. Then again, spring is when I saw my worst damage. A mouse guard is a cheap defense considering the cost of replacing the colony.

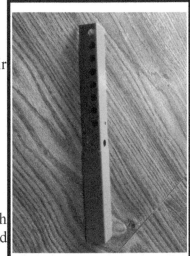

Illustration 162: Mouse guard metal bar. Photo credit: Susan Sommers

Susan ordered a metal guard with predrilled holes but was not happy with the way it fit on the entrance so she made her own. The metal one seemed to be difficult for the bees to use. Using the normal wooden entrance reducer, she nailed small nails 3/8 inch apart across the hole. The bees took to it immediately.

Apitherapy

© steve kennedy 2014

Apitherapy is the use of honeybee products (bee bread, beeswax, honey, pollen, propolis, royal jelly and venom) medicinally. The medical efficacy of some bee product treatments has not yet been proven scientifically; however, there is a large body of anecdotal evidence suggesting positive results.

Apitherapy is thought to be helpful with these medical conditions:[45]

Immune system dysfunctions or problems

- Multiple Sclerosis (MS)
- Rheumatoid arthritis
- Hay fever

Neurological problems

- MS
- ALS (Lou Gehrig's Disease)
- Shingles
- Scar pain

Musculoskeletal problems

- Arthritis[46]
- Gout
- Tendinitis, bursitis
- Spinal pain

Infectious problems

- Bacterial, viral, and fungal illnesses

Traumas

- Wounds, acute and chronic
- Sprains
- Fractures
- Burns

Tumors

- Benign
- Malignant (cancer)

Honeybees are not just "lucky" that the hive products they gather and make are medicinal. Given a specific fungal infestation, the foragers will modify their resin collecting to gather that which is specifically effective against that fungus.[47] The bees are seeking the medicinal properties intentionally. Products collected by honeybees from plants and trees are perfectly adapted to their climatic conditions. For example, plants living at high elevations secrete more bioflavonoids than plants at sea

---

45 http://www.apitherapy.org/about-apitherapy/conditions-treated/, visited 10/28/2012
46 http://www.medicinenet.com/script/main/art.asp?articlekey=50602, visited 2/2/2014
47 http://www.plosone.org/article/info%3Adoi%2F10.1371%2Fjournal.pone.0034601, visited 2/2/2014

level. This is due to higher radiation levels in the mountains.[48] As a result, honeybee products made in the mountains protect humans living in that area more-so than would the same products made at sea level. Likewise, the products made at sea level are better for inhabitants there.

This author makes no claim as to the safety or efficacy of the therapeutic use of honeybee products. However, I have noticed that since I've been keeping bees, *I am almost never sick.* Annually, I get approximately 50 stings and consume approximately 6 pounds of local raw honey.

Bee Bread

Bee bread is a mixture of bee saliva (enzymes, fungi and bacteria), plant pollen, and nectar packed into a cell by a worker using her head; it contains bioflavinoids.[49] The end product is higher in nutrition than the untreated plant pollen. It is one of the oldest dietary supplements known. Taken by some athletes to enhance performance in high-intensity exercises, claims that it increases energy levels and boosts performance by improved oxygen intake have not been supported by scientific evidence. It has *antioxidant* properties and is believed to enhance immune system function.

Beeswax

Beeswax is a biologically active product with *germicidal*, *anti-allergenic* and strong *bactericidal* properties.[50] *Anti-inflammatory* and wound healing properties make beeswax a great natural remedy for skin conditions and mucous membrane diseases. Ancient healers recommended warm wax compresses for healing.

A lipoid granular substance made by worker honeybees, the largest amount of beeswax is produced by young bees that are 12 – 18 days old. To produce 1 kg of wax, bees need to eat 3.6 kg of honey.[51]

Honey

Raw honey is good for you. The detail of why is a surprisingly long list of scientific reasons. We will explore one benefit, *honey will help you sleep*, and list others for you to explore on your own.

Honey improves blood sugar control. Fueling the liver properly is required for glucose metabolism during sleep. Honey, it turns out, is the ideal liver fuel because it contains a nearly perfect ratio of fructose to glucose.

Fructose "unlocks" the enzyme from the liver cell's nucleus that is necessary for the incorporation of glucose into glycogen (the form in which sugar is stored in the liver and muscle cells). An adequate glycogen store in the liver is essential to supply the brain with fuel when we are sleeping and during prolonged exercise.[52]

Moderate, yet regular dietary use of honey also:

- improves glucose metabolism,
- reduces insulin resistance,

---

48   http://www.apitherapy.com/index.php/eng/F.A.Q, visited 2/2/2014
49   http://www.answers.com/topic/bee-pollen, visited 11/4/2012
50   http://www.ehow.com/about_5705546_skin-benefits-beeswax.html, visited 2/2/2014
51   http://www.drchristophersherbshop.com/Beeswax.html, visited 11/4/2012
52   http://www.whfoods.com/genpage.php?tname=foodspice&dbid=96, visited 8/7/2012

- aids treatment of diabetes,
- reduces cardiovascular disease risk and
- boosts immunity.[53]

Furthermore, honey performs effectively as a cough suppressant.[54]

The ancients used raw honey as wound dressing. A gentle, *antiseptic* healing agent on ulcers and burns when applied topically as well as an *antioxidant*, raw honey contains small amounts of the same resins found in propolis. When honey is extensively processed and heated, the benefits of the phytonutrients (chemicals that help plants defend against environmental challenges) are largely eliminated.[55]

Some varietals of honey possess a large amount of friendly bacteria (6 species of lactobacilli and 4 species of bifidobacteria), which may explain mysterious therapeutic properties.[56]

When nine common pathogenic bacteria were added to honey, all the bacteria died within hours or days. Honey is not a suitable medium for certain bacteria for two reasons - acidity and high sugar content. Killing bacteria by high sugar content is called "**osmotic effect**." Some bacteria may survive osmotic effect in spore form, but they will not grow.

Pollen

Pollen is a natural *stimulant* for cell renewal. Honeybees collect pollen from local flowers. Although clinical studies confirming this practice are lacking, advocates claim that by consuming the pollen of local plants that trigger hay fever, an individual can gradually desensitize their airborne pollen reaction.[57] Based on oral desensitization, the idea is that a small amount of the allergen, absorbed directly into the blood stream, causes the body to develop a tolerance.

Propolis

Propolis is biologically active tree resin and is known to be a powerful, natural *antibiotic*, *antimicrobial* and *antiviral* agent. Honeybees make propolis by combining plant resins with their own secretions.

The resins found in propolis only represent a small part of the phytonutrients found in the many hive products produced by honey bees. Other phytonutrients found in both honey and propolis have been shown to possess cancer-preventing and anti-tumor properties. These substances include *caffeic acid, methyl caffeate, phenylethyl caffeate*, and *phenylethyl dimethylcaffeate*. Researchers have discovered that these substances prevent colon cancer in animals by shutting down activity of two enzymes, *phosphatidylinositol-specific phospholipase C* and *lipoxygenase*.[58]

Royal Jelly

Royal jelly is used for asthma, hay fever, liver disease, pancreatitis, sleep troubles (insomnia),

---

53 The Honey Revolution, Ron Fessenden, MD, MPH and Mike McInnes, MRPS, WorldClassEnterprise, 2008
54 http://www.honey.com/honey-at-home/honeys-natural-benefits/natures-cough-suppressant, visited 2/2/2014
55 http://www.nursingtimes.net/using-honey-dressings-the-practical-considerations/205144.article, visited 2/2/2014
56 http://www.lordbyronshoney.com/honey_composition_and_properties.htm, visited 8/7/2012
57 http://www.livestrong.com/article/155222-medicinal-properties-of-bee-pollen%5D, visited 11/4/2012
58 http://www.naturalnews.com/040533_propolis_cancer_prevention_immune_function.html, visited 2/2/2014

premenstrual syndrome (PMS), stomach ulcers, kidney disease, bone fractures, skin disorders, and high cholesterol. It is also used as a general health tonic, for fighting the effects of aging, and for boosting the immune system. Royal jelly contains gelatin which is essential for collagen production. There is a growing body of scientific information available about the effects of royal jelly in people. In animals, royal jelly seems to have some activity against tumors and the development of hardening of the arteries.[59]

Venom

The American Apitherapy Society, AAS, recommends that any one that practices bee venom therapy have an Epinephrine kit readily available and know how to administer it in case of an anaphylactic response.

Many people consider swelling after a sting to be an allergic reaction. This is not correct. Swelling is a normal sting response, as are localized redness and itching. Honeybee venom allergy is rare, only affecting about seven per 1000 persons.[60] A small portion of those seven risks anaphylactic shock (symptoms include: nausea, lowered blood pressure, irregular heart beat, vomiting, and difficulty breathing). Anaphylactic shock may lead to coma or death. Though rare, it must be taken seriously.

Interestingly, sting reactions may vary by bee. For instance, I generally experience some initial pain which passes within one to ten minutes (variability). I get redness and swelling that increases for two days and then declines. I can watch the swelling move through my system, for example, from my hand up through my arm. Little more than itching remains after the third day. However, I got stung in June, 2012, and immediately started feeling nauseous. This was not a multiple sting case. I got just one sting and had an immediate feeling of being about to throw up. Coincidence? Maybe. I don't know. It has not happened again despite several stings since that event.

Honeybee venom therapy, called Bee Venom Therapy or BVT, was introduced by the Austrian physician, Phillip Terc, in 1888. Bee venom is a natural *anti-inflammatory* and pain reliever. One BVT example is for the elimination of arthritis pain. The honeybee is usually held with a pair of tweezers and allowed to sting the patient at the arthritis trigger point. Normal, initial redness, swelling and pain are experienced. A single session consists of two to five stings. The patient receives the BVT treatment, sometimes every day, until the arthritis symptoms subside. At this point the patient feels significantly less pain, if any. Should arthritis symptoms reoccur, the process is repeated. BVT may also be effective for bursitis, tendinitis and dissolving scar tissue.

I experimented with BVT for my arthritic shoulder beginning in March and continuing through September. I would intentionally force a bee to sting my shoulder once or twice on a visit. The timing had everything to do with my visits to the apiary and were very inconsistent. During this six month period, I took approximately 30 stings total. I regained full mobility of my shoulder and experienced very little discomfort during the experiment.

---

59 http://www.webmd.com/vitamins-supplements/ingredientmono-503-ROYAL%20JELLY.aspx?
    activeIngredientId=503&activeIngredientName=ROYAL%20JELLY], visited 2/2/2014
60 http://www.apitherapy.org/1146/what-about-bee-sting-allergy/, visited 10/28/2012

September ... Harvest

© steve kennedy 2014

Apiculture, the practice of beekeeping to produce honey, dates back to at least 700 BC. Honey is a viscous sweetener made by bees for their own nourishment. The process begins when the bees visit flowers, collecting the nectar and, less often, honeydew in their mouths. Nectar mixes with special enzymes in the bees' saliva. The bees store the nectar in their honey crop for the ride home. There, they pass the load off to a sister who will then deposit it into comb cells. Honey is stored above and to the sides of the brood and bee bread area.

Nectar is mostly water (about 80%), a thin, easily spoiled sweet liquid. Honeybees reduce the water content by fanning the nectar, causing evaporation, until the ripe honey water content is achieved. Only after reducing the water content to 18.6% is this substance considered honey, a stable, high-density, high-energy food. At that point, the bees will cap it, preventing any further changes in water content. Nectar must not have water content higher than 19% or it will ferment. True honey has a shelf life of approximately 3000 years, give or take a few centuries.

Made, primarily, of two sugars, dextrose (glucose) and levulose (fructose), and water, honey also contains trace amounts of 22 other complex sugars and many other substances. Color varies widely from pale yellow to nearly black and may be affected by the darkening action of heat; however, the primary color determination is the floral nectar source. Darker honey contains more ash and nitrogen than lighter honey. Honey is acidic. The primary acid is gluconic acid while 17 others can also be found. Honey contains enzymes which are complex protein materials, most important are invertase, diastase, and glucose oxidase.[61]

Illustration 163: Good year. Tom Theobald. Photo credit: unknown

Harvest

Each colony needs 60 - 90 pounds of honey and bee bread to get through the cold season. The beekeeper only gets a harvest if the colony has produced beyond what they need. Honey ready for harvest is capped. Browse the beekeeper magazines and you will see a wide variety of tools and chemicals that are supposed to help you with your harvest. Most are intended to move the bees out of a super such that you can lift the super into a truck, frames still inside.

Naturally, my approach uses no chemicals. Using some extra labor at the apiary and the honey house, the beekeeper can achieve nearly the same results. This approach requires tubs that can be used as a temporary holding point for frames. A 56 quart tub can nicely hold 10 frames.

Using a flowing smoker, announce to the bees that you have arrived. Removing one frame at a time, shake the bees into the hive. The last few usually require a brush. Open the lid of the tub, insert the

---

61 http://www.beesource.com/resources/usda/honey-composition-and-properties/, visited 2/3/2014

honey frame, replace the lid. Repeat this until you have harvested all the frames of surplus honey. Tubs can be transported to the honey house nearly bee-free.

If you want to store the frames in the super box, you can easily move them from the tub back into the super at the honey house. This would free up the tub to be used at the next apiary. This is true whether you are using one tub or a dozen.

This is food we are placing in these tubs and they should be cleaned, in advance, with the appropriate amount of care.

The down side to this approach is that you may break some comb which will result in some honey leaking in the tub. Obviously, if you had just lifted the entire super into your truck, you would have avoided breaking any comb. However, being able to transport the honey nearly bee-free is a big advantage that, to me, outweighs the occasional broken comb problem.

This procedure does not address the foundationless frames you may have in a top-bar, a Warré or even in a Langstroth if you prefer natural comb. Foundationless frames are significantly more delicate. A good shake to remove the bees may well cause the entire comb to break free causing significant grief for the bees and the beekeeper. In this case, use a combination of the smoker and the brush to remove the bees over the hive. Quickly hold the comb over the tub and use your knife to cut the comb from the frame letting it drop gently into the tub. Keep the knife in the tub because the bees will be attracted to the honey on it. Replace the lid to keep the bees out. Depending on the size of your cut comb, a five gallon bucket may be a better choice than the 56 quart tub.

Picture Story

The 2012 harvest was very small, but precious. Here's some images of how honey makes it from the hives to your plate.

Illustration 164: Prepare your crush comb and spinning area for food. Photo credit: Don Studinski

Illustration 165: Tubs cleaned and dried. Photo credit: Don Studinski

Illustration 166: Arrive at the apiary. Photo credit: Don Studinski

Illustration 167: Students happy to be harvesting. Photo credit: Don Studinski

Illustration 168: Open using smoke. Photo credit: Don Studinski

Illustration 169: Next stop, short break. Photo credit: Don Studinski

Illustration 170: Working under clear blue sky. Photo credit: Don Studinski

Illustration 171: More than just harvesting, we are checking the health of every colony. Photo credit: Susan Sommers

Illustration 172: Next apiary, next day. Photo credit: Susan Sommers

Illustration 173: Obtain a yield. Permaculture principle three. Photo credit: Susan Sommers

Illustration 174: Warré diagonal comb needs to go away. Photo credit: Don Studinski

Illustration 175: Top bar, honey across the top. Photo credit: Susan Sommers

Illustration 176: Beekeeper friends. Community is part of the reward. Photo credit: Don Studinski

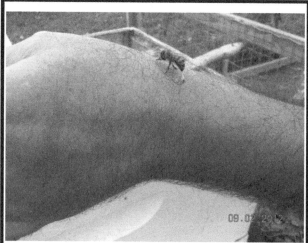

Illustration 177: A lot of harvesting. Just a few stings. Photo credit Valerie Gautreaux

Illustration 178: Solid frame of honey. Photo credit: Susan Sommers

Illustration 179: Nectar, not honey. Photo credit: Don Studinski

Illustration 180: Comb between bars is inconvenient. Photo credit: Susan Sommers

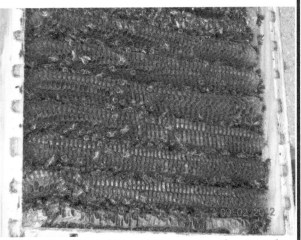

Illustration 181: Warré frames mostly straight. Photo credit: Don Studinski

Illustration 182: Next day, more of the same. Photo credit Marci Heiser

Illustration 183: Beautiful frame of honey. Photo credit: Don Studinski

Illustration 184: Newport colony built comb upward. It looks like we turned it over, but we didn't. It looks like a roller coaster. Not a good sign. Maybe they got into some poison. Photo credit: Don Studinski

Illustration 185: Superior colony produced little given the bee population. Too many bees in the area or too little nectar? We think too little nectar because of the heat. Photo credit: Don Studinski

Illustration 186: Braden girls made honey to harvest their first year. That's unusual. Photo credit: David Braden

Illustration 187: Packaging comb honey. Photo credit: David Braden

Illustration 188: Comb honey commands a premium price. Photo credit: Don Studinski

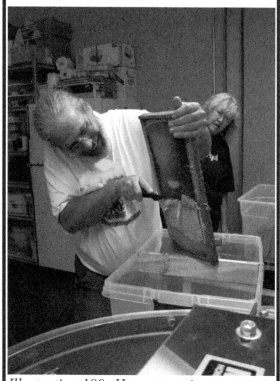

Illustration 189: Harvest cutting caps which fall in the tub. Photo credit: James Bertini

Illustration 190: Cold knife cutting pretty white cappings. Photo credit: James Bertini

Illustration 191: Load uncapped frames into extractor. Photo credit: Linda Chumbley

Illustration 192: Spin those frames. Honey flys to the sides. Photo credit: Linda Chumbley

Illustration 193: First drops hit the strainer. Pay off! Photo credit: Linda Chumbley

Illustration 194: At honey festival time, you get an audience to watch you work. Photo credit: James Bertini

Beekeeping Income

© steve kennedy 2014

P eople talk about "green" and people talk about "sustainable." In my opinion, the issue to focus on is resilience. Honey is an important food which can be provided locally by a resilient network of beekeepers. Beekeeping itself adds resilience to a community by providing meaningful work for the beekeeper and a valuable pollination service for the surrounding area.

These issues are worth pondering.

Is It Work?

Maybe you really want your beekeeping to grow to the point that it can support you. Maybe it's what you love. Maybe you want to do it full time. Then again, maybe not. Some think turning a hobby into a business turns joy into work. Others think being able to make a living doing what you love means you never have to work. Whatever your case, the fact is that beekeeping holds the potential to actually pay. Of course, it also holds the potential for severe losses. This is, after all, agriculture. For many, there can be a happy middle ground where a hobby becomes, with practice, self supporting.

Income Streams

As for me, I find it surprising how many widely different potential income streams exist under the beekeeping umbrella. There's the obvious:

1. Sell honey
2. Sell pollination service

But then, there's a long list of other opportunities, most silently lurking around the edges. It's like that permaculture principle, *Use Edges and Value the Marginal*. Let's explore some of the income opportunities that may not be obvious.

An easy split is the division between hive <u>products</u> and beekeeping <u>services</u>.

Hive products with market value include:

1. Bee bread
2. Honey
3. Pollen
4. Propolis
5. Royal jelly
6. Venom
7. Wax

Some of these have nutritional food value, but all of these have medicinal value. Learning to collect the product and prepare it for market is a unique experience for each. Pick the ones that interest you and build your expertise. Master one, then move on to the next, or not, depending on your desires.

Chances are that you have noticed nearly everyone that becomes aware that you keep bees now wants to talk with you about bees. Each of those folk is a potential customer for your honey or other products. Here's a twist, if you are good at letting people know you have honeybee products for sale, you can sell, not only your own products, but other beekeepers' products as well. Not every beekeeper is comfortable with a sales conversation. You become a wholesale purchaser from your friends and a retail seller. That friendly "talk to folks" skill can be of significant value and the cooperation with your friends can be a win-win.

Beekeeping services you might provide include:

1. Retrieving swarms
2. Public speaking
3. Teaching classes
4. Mentoring other beekeepers
5. Consulting
6. Pollination service
7. Removing colonies from structures
8. Spinning or crushing honey for other beekeepers
9. Apitherapy
10. Writing
11. Multiplying bees as livestock to sell

Many of us choose to learn to fetch swarms not long into our beekeeping career. Traditionally, this is a free service that beekeepers provide to the community. But that doesn't mean it is of no value to the beekeeper. That colony represents money not spent on a package or a nuc. It represents potentially valuable genes. It represents a mother colony that successfully wintered. Those bees may even be valuable enough to another beekeeper that they would pay you for them. Most importantly, the service is a chance to provide education to the public about the importance of bees and beekeeping. This can lead to other beekeeping-related income opportunities for you from on-lookers.

Illustration 195: Broomfield Library. Photo credit: Marci Heiser

Speaking of education, your growing knowledge about beekeeping is of interest to the general public, even those with no desire to participate in beekeeping. You can put together a slide show with photos you have taken just because you wanted them, and share it with civic clubs, local libraries, schools, dating clubs, etc. You would be surprised how easy this is to get going. Chances are your very first presentation will have someone in the audience interested to have you speak at another organization. It continues to blossom from there.

Now that you are doing public speaking about beekeeping, why not formalize it a bit and teach classes? It's actually not a large additional effort to prepare some handouts and organize a talk around the topics beekeepers need to know. There is a growing interest in urban homesteading these days with many new small-time beekeepers anxious to learn.

Every new beekeeper could benefit from a close relationship with a mentor. Some beekeeping clubs provide newbees with a mentor for free. Overall, there is a severe shortage of people willing to perform this service and a huge demand. Therefore a talented mentor can charge a reasonable fee and students are glad to have access.

Once you have developed a strong mentor relationship with newbees, you will find them requesting individualized time in addition to what you offer as part of the mentor relationship. That's called consulting and can be offered for an additional fee. For example, you may offer a mentor service that includes showing students how to confirm that a hive is ready for winter. A student may want the added service of you coming to their apiary and confirming that their hives are ready for winter. An additional hourly rate may apply.

Some folks have an interest in housing honeybees on their property, but have no interest in becoming a beekeeper. That's called pollination service. Consider offering to place hives for an annual fee. The downside is the extra driving associated with hives at many locations. The upside is yet another income stream. Having many income streams creates resilience.

Sometimes, swarms move into crevices where they are not welcome. The eaves of a home is one example. The homeowner wants those bees removed and the typical construction contractor or handyman is not willing to take on that chore due to the stinging insects. We can call this a removal or rescue. Whatever you call it, it's a service that the homeowner is willing and glad to pay for because they do not have the knowledge or experience to work among those bees.

Illustration 196: Urban Pollination Service. Sugarloaf 13 the day before the flood. Photo credit: Jaime Ngo

Illustration 197: And the day after the flood of 2013. Lower deep got wet. Photo credit: Linh Ngo

Have you grown to the point that you have access to electric honey extraction equipment? If you have, there are beekeepers that would be happy to pay you, either in money or honey, to help them with their harvest. Let your equipment pay for itself by allowing others access. You can perform the work yourself or you can rent the equipment to your customers.

Remember that all those hive products are of medicinal value. Using hive products to treat illness is called apitherapy. It's a whole separate world of products and services that can be offered. One example is people that want intentional honeybee stings to treat arthritis or MS. You might consider selling them small bottles of live honeybees which they can then use to sting themselves. In this case, the bees are free, but the customer pays a service fee for you to collect the bees for them.

Start a blog. Write about your bees and your beekeeping experiences. If you write well, someone may ask you to allow them to republish your writing. There is generally a fee.

Once you are skilled at over-wintering your bees and performing spring splits, you will find local beekeepers that are anxious to have your surplus stock. Honeybees are demanding an ever higher price as the industry continues to struggle keeping bees alive.

Illustration 198: September 11, 2013, do you think I was using smoke? Photo credit: Nancy Griffith

These are examples of products and services that can be offered by hobby beekeepers without giving up their day job. The clever person that really loves their beekeeping may be able to find a way to make this support them completely. One thing worth keeping in mind is that a person that independently earns more than their expenses for any given period of time is already independently wealthy. There are two ways to achieve this. One is to expand opportunities for income. The other is to eliminate expenses. Both can achieve a very powerful goal called freedom.

State Regulations

Every State and, potentially, every County will have unique laws regulating the sale of products and the offering of services. It is up to you to find the laws that apply to you. Sale of local raw honey falls under the Cottage Industry law in Colorado. You can find the details about this on the State of Colorado website. The law currently requires some food handling training which can be done on-line and is fairly inexpensive.

October ... Finalize Winter Preparation

© steve kennedy 2014

After the fall Equinox, it is normal for the bees to reduce their population. Brood rearing will slow.

Foraging workers will be reaching normal end of life. During the warm weather, these workers will make every effort to fly away to die. They do not wish to burden the colony with removing their body. Once the colder weather sets in, workers will find themselves home and in cluster when their end of life arrives. This is a case where the mortician bees will have to remove the body. It is not unusual to find a bunch of dead bees dropped just out the front door during fall. This is especially the case upon the first cold snaps that get near or below freezing.

Fear not. Continue to watch the front entrance on days above 50°F. There should be a few bees taking cleansing flights. If workers are hauling out the dead, this is actually a good sign. They are doing their job. The colony is functioning normally.

Cold is coming

Harvest is over.

If you have not yet verified that your girls are ready for winter, then it is time to go to the apiary. Mouse prevention should be in place. Any configuration adjustments should be completed. Supers are either off or you intentionally plan to leave them there for the cold season. Know the stores they have. Feed syrup if necessary. Perform your final heft tests. This is your last chance to see the brood area before "no peeking" begins. Plan for candy where necessary or desired.

Watch for collapses. Check your bees for activity at the front door. Warm day activity should be dwindling, but not stopped. You want to catch a dead-out as soon as possible to prevent the wax moths from taking over your precious comb. In the case of a dead-out, pull the comb out and store it where insects cannot get to it. Save it for another colony come spring or melt it to cycle it out and use the wax.

Fall is good for maintenance on empty boxes. If you have boxes needing a new coat of paint, you may want to consider swapping them out now so they are empty. If you have empty equipment, consider doing whatever repairs or standard maintenance you wish to do. It might be good to get it completed before the really cold weather gets here. Or, maybe you prefer to do that painting inside after the snow. Either way works. But be sure you hit late winter with all equipment ready to go.

If you have entrance reducers for your hives, put them in place now. Reduce that entrance so the guards have less space to defend. Wasps and other honeybees will want to rob.

If you have screened bottom boards with a slide-in "close off the screen" piece, then put that piece in place now. The bees will be glad to have a little help with that cold draft coming from below.

Use the early snow to double check your ventilation. The hive needs enough ventilation to ensure that condensation does not build up on the inner cover and cause "rain" on the cluster. At the same time the ventilation must not be so much that it causes the brood to be chilled.

Will you be selling honey next year? Have you considered business cards? Do you know the regulations surrounding beekeeping for your area? This might be a good time to work on some of this background stuff.

Do you have a list of books you have been meaning to read? Now is a good time.

If you were thinking that beekeepers spend fall and winter quietly relaxing, then I have some news for you. We do not stop. We set a pace, a reasonable life pace, and we stick to it. There IS time to enjoy life, especially in the apiary. Every day brings joy when you keep bees.

Drone Termination

Drones have not been needed for some time now because any need for mating has past long ago. Any drone left alive is now considered a liability to the colony. There is no need to feed a boy through fall and winter when a new one can be made as needed come spring. Therefore, all males are herded out of the hive and sent off to certain death. Honeybees cannot live separate from the colony; at least, not very long. Though the drones may beg to get back in, there is no friendly welcome by October. Not only are drones not welcome to come in, if they get in, the guards will gang together to drag them right back out. "Go away!" they are told, "When we want another one like you, we will make one."

A Mentor's Year

My interest in honeybees started at a community garden. I was the coordinator for a new community garden in Broomfield. As we were building our new permaculture beds, I looked around and noticed there were essentially no flying insects. Two weeks later I owned two bee hives with living colonies. I had to learn, so I did.

Now I do beekeeping full time, without other sidetracking jobs, like working for corporate America. I spent most of my adult life actually believing that the company I worked for cared about me. Now, I've been laid off and "constructively" criticized enough to know that I was living in a fantasy. Do what you love is not just another silly saying. It's an important motto to live by. Life is too short to do anything else.

January:
- Hive building (hands-on event)
- Teach Introduction to Beekeeping (classroom event)

February:
- Swarm trap building (hands-on event)
- Teach splits (classroom event)

March:
- First inspection (hands-on event)
- Three deeps
- Who to split
- Hive cleanup
- Equinox celebration

April:
- Splits / queen cell checks (hands-on event)
- Swarm collection begins (hands-on event)

May:
- Confirm queen health (hands-on event)
- Merge failed splits (hands-on event)
- Swarm Collection (hands-on event)

June:

- Assess harvest (hands-on event)
- Solstice celebration

July:
- Mite control (hands-on event)
- Queen cell checks (hands-on event)
- Merge failed splits (hands-on event)

August:
- Harvest preparation (hands-on event)
- Assess winter stores (hands-on event)

September:
- Harvest (hands-on event)
- Sales
- Equinox Celebration

October:
- Equipment modifications / maintenance (hands-on event)

November:
- Make candy supers (hands-on event)
- Make candy (hands-on event)

December:
- Feed candy (hands-on event)
- Teach pesticides (classroom event)
- Solstice Celebration

Eliminating a Laying Worker

© steve kennedy 2014

A few workers are always laying eggs, even in a queen-right colony. Given the healthy pheromone levels of a normal queen and her brood, the worker's eggs are quickly cleaned up by other workers as part of their normal housekeeping duties and these laying workers do not gain royalty status.

Once your hive has been queenless for three weeks, it may develop laying workers that gain queen-like loyalty from their sisters. This is commonly called a laying worker or drone layer, though there may be several workers actually doing the laying. Workers are not capable of having sex and cannot carry the sperm necessary to produce fertile eggs. Any egg laid by a worker will be haploid, not fertile, destined to become a drone. Drones do not gather nectar or pollen and do not contribute to the day-to-day survival of the colony. So, a laying worker colony is destined to perish.

Caught early enough, you might be able to get some help from the remaining worker population to transition to a new queen. That's what we will hope for. But, a queen is a colony and a colony is a queen. Now that the queen is gone, the colony is lost. All that's left is their slow demise via attrition. This is true even if you do successfully introduce a new queen. Her offspring represent a new line of genes. The old line of genes terminates when the last of the missing queen's daughters are dead.

Beekeepers have several procedures which can be used in the laying worker case to help with the transition to a new queen. Simply introducing a new queen in the standard fashion will almost always result in the new queen being killed. This is because the workers have developed a loyalty to the laying worker(s) just as though she was a queen. Introducing a new queen is not enough to break that loyalty issue.

What follows are some options for you to consider in your response to a laying worker.

Shake and Forget

A queen is a colony and a colony is a queen. She is gone. You have lost them. Free up your equipment and move on. Shake them into the weeds at least 100 yards away from the original position. Don't put the hive back where it was. Force the foragers to beg their way into another hive or die.[62]

Give Them Brood

Take a frame of eggs, larvae and brood from a strong colony and introduce it to your laying worker colony. Check it in a week to see if they are making queen cells. If yes, let them do what they are doing. If no, get another frame of eggs, larvae and brood from a strong colony and introduce that one to your laying worker colony. Give them another week. Check again for queen cells. If none, try it one more time. They will generally have produced queen cells by the third try.[63]

Failing this, it's time to shake and forget.

---

62 http://www.bushfarms.com/beeslayingworkers.htm, visited 2/9/2014
63 http://www.bushfarms.com/beeslayingworkers.htm, visited 2/9/2014

Queen at the Wall

Remove a frame or two of brood with open larvae and plenty of bees from a strong colony. Make sure you have the queen on one of those frames. Place these frames into the troubled hive next to the hive wall. The queen should be on the wall side of the frame next to the hive wall. This will give her protection from workers loyal to the laying worker(s) who will want to kill the real queen. Expect some fighting along the boundary between workers loyal to the real queen and workers loyal to the laying worker(s). Over time, pheromones will spread and the living workers will all develop a preference for the real queen. The laying worker(s) will revert back to their previous role. The colony is once again queen-right.[64]

The colony you stole that queen from is now queenless. You should requeen them using your favorite method. There is no laying worker issue there.

Queen in a Cage

You can create a cage using hardware cloth. This is a rectangle with six sides, the one side is open. You place the queen in this cage and push the edges into some empty brood cells where she can lay. It's important that some of the cells under the cage have emerging workers. They will be the nurse bees to care for the new eggs and larvae the new queen is about to provide. As the new queen pheromone spreads and as her brood pheromone spreads, the loyalty of the remaining workers will begin to shift from the laying worker to the new queen. The laying worker(s) will probably not be eliminated, but they will revert to the role they had before the colony was queenless. Let her stay under the cage for about a week. Observe that the workers are no longer biting on the cage angrily before you allow the queen to escape the cage.[65]

Extended Merge

Using a double screen to separate the two colonies, temporarily place the troubled hive on top of a queen-right hive and leave them to get used to the mixed pheromones from the real queen and the laying worker(s) for two or three weeks. Then perform a normal newspaper merge to finish the job.[66]

Notice that in this case, you have spent quite a bit of time and still are down one colony in the end.

Shake and Hope

This approach has the built in assumption that the laying worker(s) have not flown outside the hive, that is, they are not oriented to where the hive is. Therefore, if you displace her, she cannot find her way home.

Pick up the entire hive and walk it at least 50 yards, better yet 100 yards away from its original position. Dump and shake all the bees out of the hive and off of the frames. Yes, I mean all the bees. You must get the laying worker off and you don't know which one she is. Reassemble the empty equipment back in the original position. The foragers will quickly find their way home, but the new bees that have not performed orientation flights will be left in the grass. This includes the laying

---

64   http://basicbeekeeping.blogspot.com/2010/07/lesson-75-dreaded-laying-worker.html, visited 2/9/2014
65   http://basicbeekeeping.blogspot.com/2010/07/lesson-75-dreaded-laying-worker.html, visited 2/9/2014
66   http://www.bushfarms.com/beeslayingworkers.htm, visited 2/9/2014

worker(s).[67,68]

At this point the population left in the hive is queenless. That is, they don't have a loyalty issue to the laying worker(s) which have been removed. In a few hours, you can begin a normal requeening procedure and within a few days they will accept the new royalty.

My attempts at this procedure have not been successful. Perhaps I didn't shake them far enough or perhaps the laying worker(s) can find their way home. This is not clear.

67 http://www.dummies.com/how-to/content/how-to-get-rid-of-laying-workers-in-your-beehive.html, visited 2/9/2014
68 http://www.dave-cushman.net/bee/layingworkers.html, visited 2/9/2014

November ... No Peeking Begins

© steve kennedy 2014

Beekeepers would, ideally, like to be able to winter our bees without supplemental feeding. Bees, after all, have been getting through winter far longer than humans have been managing bees. Bees, planning ahead, store honey and bee bread specifically for this purpose. These days, with winter losses consistently hitting 30-40%, many of us are turning to feeding as a way of increasing our chances of getting to spring with live bees. Where Old Man Winter can keep

Illustration 199: Feeding jar, one gallon. Photo credit: Don Studinski

temperatures down in the 20s°F or below for extended periods of time, it's nice to have a way to get supplemental feeding to your bees without dealing with liquid syrup feeders. Liquid feeders, especially in cold temperatures, can potentially do harm by chilling your bees, which is clearly not what you set out to do when you decided to feed them.

Illustration 200: Feeder, plastic bucket. Photo credit: Don Studinski

Keeping bees in the Denver metropolitan area calls for 60 to 90 pounds of honey and bee bread per colony to get through the cold season. Here, the cold season is the latter part of fall and nearly all of winter. I make sure that my harvest, if any, leaves adequate stores of natural food for my girls. However, I have had the unfortunate experience of opening a colony in spring to find all the girls head down, butts in the air, dead. This experience has sensitized me to want an insurance policy. I call this insurance "bee candy."

I've tried syrup feeding in the fall, but found that to be too labor intensive, standing by the stove nearly every night to ensure all the feeder jars are topped off early the next day. I found that this ritual had to be repeated for several weeks to get adequate stores prepared. I also didn't like this solution because I was always vulnerable to a mason jar full of syrup freezing and breaking, giving me a mess and leaving the bees without food. Or worse yet, setting off a major robbing frenzy which results in a lot of dead bees.

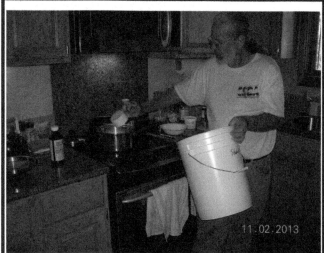

Illustration 201: Five gallon bucket is full of sugar. Photo credit: Nancy Griffith

"No peeking" is an issue unique to climate which can vary from year to year. In Colorado, beekeepers should plan for no peeking to begin November 1 and run through middle of March. Every rule has exceptions. Things will come up that require or tempt you to look. The one exception that I can always count on is putting on hard candy with protein after the solstice. I don't do this for every colony, but I always seem to have colonies that need it. Or, maybe it's me that needs it as "insurance" for that colony.

Still, I plan to be ready for "no peeking" by November. All preparations for the cold season are complete.

It might seem like a beekeeper gets a break for off-season. Not so for the beekeepers I know. This is time for equipment repairs, teaching classes, new equipment assembly, attending conferences, reading books and all the other indoor activities that can't be accomplished during the warm season.

Feeding Honeybees

Winter is the only time I consistently find myself with colonies that need me to (or I want to) feed. But while we are thinking about and preparing for feeding, it's a fine time to discuss the topic generically and consider both motivation and appropriate food for each season throughout the year.

Fact:

Bees, like all living animals, need both carbohydrates and protein. In nature, the honeybee carbohydrate is nectar or honey and the protein is bee bread which is made from pollen. Humans have found convenient substitutes for both which can be used to feed honeybees. Which substitutes you choose, or if you choose any at all, is a decision you get to make and a responsibility you must bear.

Illustration 202: Candy boiling. Bee tea on back burner Photo credit: Eileen Callaway

Opinion:

First, consider carbohydrates. In spring, March 22 – June 21, the motivation to feed would be to stimulate wax glands to ensure the girls are able to expand comb as they wish; therefore, I would use 1:1 sugar syrup with a bit of Honey-B-Healthy® (HBH) used according to label directions. Nectar is, or soon will be, abundant, so most of the syrup will be consumed by the now-living bees and if they choose to store some, they have plenty of time to fan it down to reduce the moisture content. The most likely colony to get fed by me during spring would be a newly installed package.

Nectar dearth in summer, June 22 – September 21, may also be a motivation for feeding. Adjust your recipe according to heat and humidity during this time. High heat or low humidity calls for more water. But, if you are hoping for a honey harvest, you should not be feeding the bees, you are hoping they will be feeding you. Keep in mind that sugar syrup capped in cells is not honey. Therefore, if you choose to feed, then you must not have supers on which you intend to use for harvest.

Illustration 203: Watching the thermometer can be challenging for old eyes. Photo credit: Eileen Callaway

In fall, September 22 – December 21, the feeding motivation is to finish over-wintering preparations; therefore, I would use 2:1 sugar syrup, again with a bit of HBH. Nectar is getting scarce and most of the syrup will likely be stored in cells. Time is short to reduce the moisture content and using a thicker syrup gives the bees a bit of help. In both cases (spring and fall) the ratio is sugar:water. You can measure by weight or by volume, this is not an exact science and your choice is not significant to the bees.

Illustration 204: David Braden designed and made candy forms. This one shows wax paper ready to pour. The bricks come out 10.5" x 7" x 1". If you let it get too hard in the form, then it can be quite difficult to get out (not that I've ever done this). If you pull it out while it is still a little warm and soft, then it comes out very easily. Photo credit: Eileen Callaway

Here, we take a short sidetrack to address protein. Feeding protein (pollen paddies, or equivalent) is motivated by a desire to stimulate brood rearing. As winter wanes, bees should be building up their population such that they have a large workforce ready when the day comes to harvest spring's abundant nectar flow. Not so in fall. A declining population in fall is normal because the colony needs to limit the number of mouths to feed during the cold weather. That said, there is a limit where declining population may be a bad thing. Whenever the cluster size falls below the population necessary to maintain temperature (and this will vary as ambient temperature varies), the entire colony will perish. This motivates some beekeepers to feed pollen paddies in fall. This is especially true if the bees will be traveling to southern California for almond pollination in February.

Winter, December 22 – March 21, presents temperature challenges for feeding. Liquids can freeze and, at a minimum, they act as a thermal mass that retains cold. This can actually be a detriment to the honeybees when the liquid is inside the hive. Liquids outside the hive have a reputation for stimulating robbing behavior and cold weather season is a time when the colony is least able to defend against robbing. This is one good reason beekeepers frequently reduce entrances when the weather starts to turn and leave them reduced until nectar is again abundant.

Illustration 205: Candy super on a queen excluder as an example of where to load a candy brick. Photo credit: Eileen Callaway

My alternative for winter feeding is hard candy. My recipe includes protein, therefore it is not recommended for before winter solstice (Dec 21). However, after solstice, I want to stimulate brood rearing and I have been pleased with the results using this technique. Placing a brick of hard candy over the brood and inside what I call a "candy super" allows the warmth of the cluster to rise, soften the candy, and be just right for a bite from a hungry worker. The idea is to provide the carbohydrate and protein in a single, non freezable source. You can see the recipe and pictures of the end result in the next chapter.

I prefer that bees eat what bees have stored for themselves to get through the cold season. The bees are much smarter than I am about what they need. Still, sometimes I like a little "insurance" to increase my likelihood of arriving at spring with a healthy, thriving colony. That's when I choose to feed. I feed bees in an effort to save those that have shown good productive attributes and earned the right to be saved from unusual circumstances. Others, I do not feed.

Keep in mind that we all buy insurance in different amounts at different times for different motivations. The same is true for bee candy. For example, my 1988 Toyota Beemobile is old and rusty, therefore, I have the required liability coverage, but I do not have collision or comprehensive on that vehicle. On the other hand, the 1999 Honda Accord is in good enough shape that I want the collision and comprehensive coverage. I really can't afford the repairs should it be wrecked. You are all making similar decisions about your insurance coverage.

Illustration 206: Filling the form with candy. Notice the candy is brown from the MegaBee. Also notice that the form doesn't quite fill up. Photo credit: Eileen Callaway

Likewise, I have colonies for which I do not need insurance coverage (candy). All three Birch colonies are examples. The mother Birch colony produced honey in 2012. This second-year colony is vulnerable to a crash due to mites, but they have all the honey and bee bread they need to get through winter. They will get no candy. If they make it, I'm happy to have them for a third year, but if they don't make it, I'm not going to be surprised. It's my anticipation of their death due to mites that motivated me to do the three way split that produced the daughter colonies. I wanted to propagate the genes. Both first year daughter colonies, Birch 2 and Birch 3, have plenty of stores for winter. Again, they will get no candy. All three colonies must make it without intervention from me. If they die, then I was wrong about the genes. That indicates genes I do not want. Rather, I want the equipment freed up for good genes.

The place to use the candy insurance is on a colony that has genes you want, but they might be a bit short on stores. Consider the Quay girls. This first-year colony represents bees that know how to get through a Colorado winter. How do I know that? Because they are a spring swarm. Their mother colony came through the winter of 2011/2012 strong enough to throw a swarm. The Quay girls are that swarm. If they make it through the winter of 2012/2013, then they represent bees that know how to get through two Colorado winters. That's genes I want! But, they are a bit light on stores. When we checked them for harvest in September, we had nothing for humans and, even leaving everything for them, the hive was a bit light. This is the case where I want candy insurance. I'll put a brick of candy on them around Christmas or New Year.

The timing is important. My candy includes protein, the MegaBee™, in the recipe. That protein will stimulate brood rearing. If I stimulate brood rearing too early, the population of bees may expand too early and the winter stores may not last because there are too many mouths to feed. The candy is only intended to fill in the small gap that may exist after the winter solstice, but before first inspection in March. Their 2012 stores must get them through to at least the winter solstice. If not, then, once again, these are genes I do not want. By first inspection in March, they once again have access to pollen and nectar naturally. The candy is only for that small amount of time between.

Illustration 207: MegaBee for candy. This protein substitute stimulates brood rearing. Photo credit: Marci Heiser

When I open up the hive for first inspection, if the candy is still there, then I know it was unnecessary. Great! Now, I like those genes even more. I'll take that candy back and save it for the next year. It stores fine with no refrigeration. On the other hand, if the candy is gone, then I know it was absolutely necessary. That insurance paid off. As a result of the candy insurance, I have living bees that probably would not have made it otherwise.

Illustration 208: Making bee tea as a water substitute. Photo credit: Nancy Griffith

I want honeybees that know how to make it right here where I live in the time that I'm living. I don't want to interfere with nature's mission of choosing the right bees to survive. This means that I, generally, do not feed my bees. Only after careful consideration of specific circumstances will I make an exception. But there are exceptions, almost always.

Can you think of another appropriate exception that we considered earlier in the year? Hint: this is the case for feeding sugar syrup, not candy. When might the girls find themselves needing full tummies, but not having an opportunity to prepare in advance? One answer is a package. Bees shaken out of their hive and combined with other bees from who knows how many colonies and then a unrelated queen shoved in for them to "get acquainted." Did they have time to prepare for this with full tummies? No. Is feeding a must? No. If the flow is on where you are installing them, why feed? On the other hand, if there is a nectar dearth, then feeding may be necessary. This is a case for syrup:water, 1:1. Another answer is a structural removal where natural stores get damaged during the removal operation. The bees have been ripped out of a wall somewhere. No advanced warning. No time to prepare. They almost definitely need feeding. If it's before the summer solstice, use 1:1 syrup. If it's after the summer solstice, use 2:1 syrup.

You are the beekeeper. You must think about their needs. This is affected by their circumstances both present and recent past.

Let's Celebrate Solstice

Solstice and equinox are going to be important to your future as a beekeeper. The bees operate on a very meaningful calendar, and we must adjust to their timing. It is in that spirit that we, the Beekeepers, gather four times through the year, just to visit. Get to know your fellow beekeepers. These wonderful human beings that share your passion for the honeybees. This doesn't have to be a fancy gathering. No dressing up. No gifts. It's not about holidays. It's about honeybees. Enjoy a visit with your beekeeping community.

Winter solstice is 12/21. This is the day with the least amount of sunshine all year in the Northern Hemisphere. This day marks an important transition in the lives of those honeybees living in cluster in our hives. The bees will start expanding their population again. From equinox to winter solstice, the hive mind ramps down brood production. The brood nest ends up being very small, about the size of a golf ball. Solstice is the transition to begin increasing brood production again in anticipation of the spring nectar flow. Now they need and want extra protein to feed the brood. That's why we put MegaBee in the candy we put on after solstice.

Now is the time to plan your celebration and get it on your friends' calendars before the holiday season takes over.

Good Time for Laundry

Sunny days and low humidity remind me to do my beekeeper laundry.

If you are like me, you keep your suit, gloves etc. in the car all the time. That stuff almost never makes it to the laundry room inside. I wear mine quite a bit in the summer and even though I have other clothes on underneath, it's still a bit ripe by the end of the season. I wash my stuff in the bathtub by hand. I tried it in the washer once, but the netting got ruined. I hang it out to dry in the sun. That's what got me thinking about this today.

Another thing to consider while we are talking about laundry is that whenever you have an encounter with your bees that results in a lot of stings to your clothing, you should consider washing before you use those clothes again at the apiary. All that venom in the cloth will be noticed by the bees if you don't wash it out. That smell will cause them to be more aggressive during future work.

Diligent and effective use of your smoker will consistently prevent hostile encounters with your bees. So, you can avoid loading up your jacket and gloves with venom if you choose. However, the day will probably come when you decide to pop the top without smoke. After all, you will just be in there for a few seconds. Sometimes you will get away with this just fine. But, someday, those little darlings are going to be hostile and one second after you pop the top, 100 or more bees will be on you stinging.

So, the main point is: wash the venom out of your beekeeper's suit and hang it outside to dry. Pick a slow day with sunny weather for this task.

Check In On Your Bees Regularly

When the weather looks reasonable for bees to fly and poop, that is, near or over 50°F, get on out to your bee yard and have a look at the front door. There should be one or two bees at least. You may have to give it a few minutes to observe. By doing this regularly, you will have a good sense of when something has changed. Should you see no activity at all, or robbers that don't look like this is their home, then you may have a dead colony. If you check two or three times and there is consistently no activity, then it may be appropriate to look inside. Knowing the timing of a death can help you understand what may have been the cause.

If you find a dead-out, you should perform a postmortem. You want to know why they died. Was it starvation? Was it just a very low temperature for which the cluster was too small? Each of these has specific symptoms to look for in your dead hive.

Student Questions

Sandra writes:
My bees are still very active on nice days - yay! I have two questions for you:
I bought mouse guards instead of making them, just never got to it. The holes are partially covered by the front of the hive body. The bees can get in and out, but they have to wiggle and work at it a little. I think there is not enough space between my bottom boards and the hive body - must be 3/8". Is this going to be ok for the winter? My bottom boards are the same depth on both sides too.

Also, I put Popsicle sticks under the inner cover. Just one on each corner - is that right? Seems like very little extra air flow. Do I need more than one high Popsicle sticks?

Thanks, I am so anxious to have at least a couple of my hives survive the winter.

Answer:

Your mouse guard is making it hard for the bees to enter. If they are carrying a pollen load, it may fall off their legs. Consider making the opening just a bit larger. But what if we get a deep snow, like three feet. Will that not completely lock them in and block ventilation? Are you prepared to go out in the snow to clear the front door for them? They must be able to get air.

My inner covers generally have a notch for ventilation and an upper entrance. An upper entrance can be an advantage just in case the bees face three feet of snow sometime during the winter.

Popsicle sticks is an interesting idea. I don't think it's a problem as long as it doesn't leave a draft making them cold. I'm more concerned about it being too much ventilation than I am about it being not enough. Still, it's a small crack and, if they want it closed, then they will seal it up with propolis.

Sandra:

Deep snow would be an issue. I am fine to clear them this year. I don't want to drill a new entrance at this point. I should be good for winter then. My fingers are crossed!

Don writes:

Today, 11/21/2012, I drove around to my apiaries and watched front door activity. I am amazed to report that all 24 of 24 are still alive. Needless to say, I am one happy camper.

1. Birch, 2011 swarm, alive, busy
2. Birch 2, a 2012 daughter, alive, busy
3. Birch 3, a 2012 daughter, alive
4. Superior, 2010 swarm, alive, busy
5. QE II, 2012 daughter, alive, will get candy after solstice
6. Newport, 2012 swarm, alive, will get candy after solstice
7. Braden, TX nuc, alive, busy
8, Wiley, 2012 swarm, alive, will get candy after solstice
9. Rinehart, TX nuc, alive
10. Sable, 2012 swarm, alive, will get candy after solstice
11. Oak, 2012 swarm, alive, need candy, but will not get it due to experiment in progress
12. Quay, 2012 swarm, barely alive, will get candy after solstice
13. Cottonwood, alive
14. Pumpkin 1, alive, busy
15. Pumpkin 2, alive, busy
16. Arapahoe, 2012 swarm, alive
17. Clarkson, 2012 swarm, alive
18. Cherokee, 2012 swarm, alive (merged with Quay 2, another 2012 swarm)
19. Violet, 2012 swarm, alive, busy, will get candy after solstice
20. Sugarloaf, 2012 daughter, alive, will get candy after solstice
21. Violet 2, 2012 swarm, alive
22. Osceola, 2012 swarm, barely alive, will get candy after solstice
23. Pumpkin 1.1, 2012 daughter, alive, will get candy after solstice
24. Franklin, 2012 swarm, alive, will get candy after solstice

Each colony that will get candy has some special attribute that motivates me to feed them, as insurance, to increase my chances of continuing that line. Those that don't get candy are either heavy with honey and bee bread or they must make it on their own, without aid. As a general rule, I do not feed. For example, I did not feed syrup, at all, summer or fall of 2012, despite drought and extended periods of 100+ degree temperatures. As many of you know, I only harvested off of eight colonies this year and I only got 90 pounds of honey.

Honeybee Candy

© steve kennedy 2014

"**B**ee candy" is a nice dry solution that I can use in the coldest part of winter. I hope the bees won't need it, but it feels good introducing that bit of insurance that helps me sleep better. I originally got the idea of hard candy for bees from Mel Disselkoen's website[69]. Then, not knowing what I was doing, I got my first candy cooking lesson from my friend and local beekeeper, Denise O'Connor. Since that time, I've become bold and modified the recipe to the point that I'm very pleased with it. I'm hoping you will find it beneficial too.

This is a picture of Mel opening up a hive in spring that has been fed bee candy. Mel's candy doesn't look exactly like mine (it's white), however, this gives you a good idea of what you want to see after the girls have been feeding on the candy. *Thanks to Mel Disselkoen for providing the picture of the bee-candy residue left in spring after the girls have been enjoying it.*

Illustration 209: Spring candy super. Photo credit: Mel Disselkoen

Glucose vs HFCS

Glucose keeps the candy a little soft. I find my glucose at Hobby Lobby(TM) in the cake decoration section; a one-cup container (look for purple label) makes four bricks.

Illustration 210: Glucose. Photo credit: Don Studinski

Although similar in taste, glucose and High Fructose Corn Syrup (HFCS) are chemically different. Naturally occurring in nature, glucose is a monosaccharide. HFCS is created, starting with glucose or corn syrup and adding an enzyme called invertase. This causes a chemical reaction turning half the glucose molecules into a sweeter form of sugar called fructose.[70]

Do not use high fructose corn syrup (HFCS) as a glucose substitute. When heated, HFCS creates the compound hydroxymethylfurfural (HMF). HMF is toxic to bees.[71] It does not take much heat, as a dramatic increase in HMF occurs at 120°F.[72]

---

69 http://www.mdasplitter.com/, visited 2010

70 http://www.livestrong.com/article/444047-what-is-the-difference-between-glucose-corn-syrup, visited 2/10/2014

71 http://en.wikipedia.org/wiki/Hydroxymethylfurfural, visited 2/10/2014

72 http://www.sciencedaily.com/releases/2009/08/090826110118.htm, visited 2/10/2014

Sugar

Do not use GMO sugar because it will likely contain neonicotinoid residue, systemic poison which can kill your bees at 3.6 ppb (parts per billion). For example, corn and beets are usually neonic crops. Use Pure Cane sugar.

Recipe and Advice

- 4 cups pure cane sugar, about 2 lbs
- 1 cup water
- 1/4 teaspoon vinegar
- 1/4 cup glucose
- 1/2 cup MegaBee for protein

Tools you need:

- 4-quart pan
- Whisk
- Butter knife
- Spoon
- Spatula
- Measuring spoons
- Measuring cups
- Hot pads
- Parchment paper
- Candy thermometer

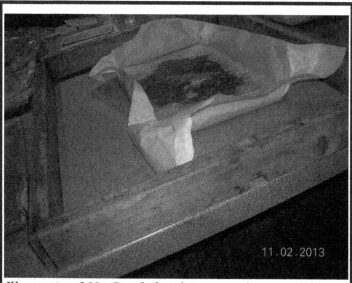

Illustration 211: Candy brick in a candy super. Photo credit: Don Studinski

Illustration 212: Candy at fudge consistency (234°F). Photo credit: Bill Koeppen

Steps:

1. Place wax paper or parchment paper on counter top or on a form if you wish.
2. Add 1/2 cup of hot water to the 1/2 cup MegaBee powder and stir to make a paste. This will allow the powder to blend right in later. Set this to the side with a spatula ready.
3. Boil sugar, remaining 1/2 cup water, vinegar, and glucose to 236°F (firm ball), stirring with a whisk. Watch the candy thermometer. I noticed that it rises, then pauses, then rises rapidly to the desired temperature. Don't let that pause fake you into looking away. You have to hit 236°F exactly or slightly above for my altitude (one mile high). I like the results at 236°F.
4. Remove from heat and quickly whisk in the 1/2 cup MegaBee, which turns the candy brown. At this point, your candy is starting to harden, and if you dilly dally, it will be too hard to spread before you know it. Try to get the MegaBee mixed in within 2 minutes.

Options:

As a water substitute, you can make bee tea[73]:

- 1 quart of water
- chamomile tea (two tea bags)
- 1 teaspoon Honey B Healthy(TM)
- 1/2 teaspoon of natural sea salt with minerals (typically not pure white in color)

Using this water substitute will cause the candy to bubble up a bit more than it otherwise would. Just turn the heat down a bit until the bubble-up stops (this happens as the liquid starts to look clear) and then turn the heat back up.

One pound of sugar is about equal to 2 cups of sugar. So, for this recipe, I found it convenient to use a 4 pound bag of sugar.

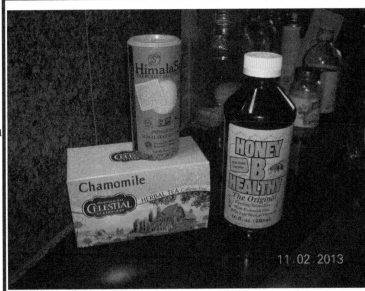

Illustration 213: Bee tea ingredients. Photo credit: Don Studinski

First batch you must measure out the 4 cups (2 pounds); next batch you can just dump in the remainder of the bag.

Illustration 214: Candy pouring into parchment paper covering the form. Photo credit: Marci Heiser

We have a great time making bee candy around Thanksgiving. A bunch of beekeeping friends get together to exchange tips and help make the candy. This ensures the candy bricks are ready to go when we get a day where the temperature spikes above 50°F *after solstice* (last day of fall, first day of winter). This recipe includes protein, the MegaBee, and will stimulate brood rearing. I do not recommend feeding protein before solstice. If the bees create too many mouths to feed, they may starve unnecessarily.

Placement:

In the case of a Langstroth hive, I put the candy brick directly above the brood area. I have a special "candy super" that will essentially fit the candy brick with some bee space around the candy. The inner cover goes directly on top of this little super, and the telescoping cover goes over that as usual. This works out nicely because the heat and moisture from the cluster rises and hits the hard candy, making it just a little bit soft and just right for a bee to take a bite. This is the same for a Warré hive.

In the case of a top-bar hive (TBH), I put the candy brick in the hive opposite the brood nest end. I have to break the candy brick in half to get it to fit. In my TBH, the brood nest is at one end, near the front door. Then there's all the honey expanding toward the other end. I've saved some room at the far other end using a divider board in the TBH. When winter candy time comes around, I move the divider board closer to the opposite end, giving me some space for the candy brick. Break it in half and put it in there. They will find it if they need it.

In this picture you can see a brick cooling in the foreground. It gives you a sense of the dimensions of a completed brick.

Here's a picture of a top-bar with space saved to the left for candy when the time comes for winter feeding. The brood nest is to the far right.

Illustration 215: Completed brick next to form covered with wax paper. Photo credit: Eileen Callaway

Illustration 216: Top-bar hive with candy space reserved to the left. Photo credit: Marci Heiser

December ... Candy Installation

© steve kennedy 2014

December is our one potential peek inside the hive during winter. In Colorado, the weather will almost always provide a day sometime after the solstice and before long into the new year where the temperature spikes above 50ºF. That's our day. Not every hive will be opened, only those we want to have candy. That is unless we have an exception come up like we had in January of 2012. Opening the hive in December should be very quick. We don't want to unnecessarily chill the cluster.

Planning Winter Chores

On the positive side: Wow, all 24 colonies still alive 12/1/2012.

On the negative side: We have not had a drop of water for 2 months. It's much too warm compared to the past. What is "normal" moving forward? We don't know.

I'm thinking heavily about April, 2013. If you have living colonies, you will want to think through these issues as well. Get a plan. You might have to prepare and move equipment, make new stands, get a new apiary started, or all of these.

For any given colony, if they are expected to hit April strong, then what is your plan for the swarm that may be created? Will you prevent it? Will you split the colony? How many new colonies can be created? Answering these questions will help you plan for how much additional equipment you need and where you need that equipment to be by spring.

No, you don't get a break. There are hives to build and paint and position for the coming season. There are supers to prepare and position for the coming season.

This is how 2012 is ending. There is a lot to do to get ready for spring.

1. Mama Birch, may be able to be split three ways requiring two more Langstroth hives.
2. Birch 2, a daughter queen, may be able to be split three

Illustration 217: Can you tell where the cluster is located? Photo credit: Molly Turner

ways requiring two more Langstroth hives.

3. Birch 3, a daughter queen, may be able to be split three ways requiring two more Langstroth hives.
4. Superior, may be able to be split three ways requiring two more Langstroth hives.
5. QE II, may be able to be split two ways requiring one more Langstroth hive.
6. Newport, may be able to be split two ways requiring one more Warré hive.
7. Braden, may be able to be split three ways requiring two more Langstroth hives.
8. Wiley, may be able to be split two ways requiring one more Warré hive.
9. Rinehart, may be able to be split two ways requiring one more Langstroth hive.
10. Sable, top-bar hive, let them swarm, arrange for swarm trap in advance.
11. Oak, Warré, let them swarm, arrange for swarm trap in advance.
12. Quay, may be able to be split two ways requiring one more Warré hive.
13. Cottonwood, may be able to be split three ways requiring two more Langstroth hives.
14. Pumpkin 1, may be able to be split three ways requiring two more Langstroth hives.
15. Pumpkin 2, may be able to be split three ways requiring two more Langstroth hives.
16. Arapahoe, may be able to be split two ways requiring one more Langstroth hive.
17. Clarkson, consider split from top-bar hive to Langstroth requiring one more Langstroth hive.
18. Cherokee, may be able to be split two ways requiring one more Langstroth hive.
19. Violet, may be able to be split two ways requiring one more Langstroth hive.
20. Sugarloaf, may be able to be split three ways requiring two more Langstroth hives.
21. Violet 2, top-bar hive, let them swarm, arrange for swarm trap in advance.
22. Osceola, Warré, let them swarm, arrange for swarm trap in advance.
23. Pumpkin 1.1, daughter queen, may be able to be split two ways requiring one more Langstroth hive.
24. Franklin, consider split from Warré to Langstroth requiring one more Langstroth hive.
25. Bill's Queen Street, may be able to be split two ways requiring one more Langstroth hive.

If this played out exactly this way, then I would need 27 Lang hives to receive new colonies, three Warré hives to receive new colonies and four swarm traps to capture new colonies. Then, I have to place all 34 new colonies geographically. This is before I've captured any new swarms. Swarms could add another 15.

So, this gives me some new tasks which need to be completed before March when things will start getting busy again. For example, I need to move Superior to make room for the Braden split. I need to make more room for a split near QE II and Rinehart. I need to make more room for hives at my Cottonwood Farm apiary, the Rock Creek apiary and the Hygiene apiary. I also need to prepare four swarm traps and get them in place. This sounds like a pretty busy winter to me.

The point is that you will need to plan out your spring well in advance so that you have time to get the required preparations done during winter.

This is a plan! The actual came out much differently. But this gives you the idea of what you should be thinking about now for your colonies. What is your plan for splits come spring?

Inconvenient Interruptions

Moving honeybees in winter is highly unusual. Normally, I would just say, "No." However, "normal" frequently doesn't happen. This is going to be true even more often as we move into the global warming age.

Planning spring splits has caused me to notice that I need the Superior girls moved from LSI to Cottonwood Farm. Why? Because I need the room next to the Braden girls for when we perform that split. Superior is sitting in the way. That's a 40 mile trek, LSI to Cottonwood. Because Superior is a three deep hive, it stands 34-inches tall. Luckily, David Braden's truck has a topper with an opening of 37-inches. We can just fit. So, we can move the hive inside the topper and not expose it to the wind during the move. I do not want to reduce to two deeps at this time because:

I can't be sure they have already moved up enough to leave the bottom deep empty (this will be the case by mid-March). and I don't want to break any more of their propolis seal than I have to. Yes, I know moving it will likely break the seals, but I'm not going to start separating deeps this time of year.

Illustration 218: Preparing to move Superior with a helper. Photo credit: David Braden

This opportunity to move is weather dependent. It must be near 50°F. If it's much colder than that, then the bumping around will kill bees when they break cluster. If it's much warmer than that, then there will be bees out flying and they will be lost. Therefore, we can't set a date very far ahead. I'll be watching the weather and when nature says "go" then we will roll on this.

But I can't do this alone. I'm going to need help. That three deep probably weighs about 200 pounds at this time of year. I cannot lift that into and out of the truck without help. So, I'll need to get help lined up in advance.

I'm pretty sure the Osceola girls are dead. I've watched them for a few warm days and I'm not seeing any bees flying in or out. We need to perform a postmortem on that colony and analyze why they died. This doesn't have to happen on any particular day, not even in December. We could even move this inside and perform the task on a cold day.

Student Questions

Rosemary writes:

I think I have lost one of my hives. The bees were okay last week but I have not seen any activity over the last few days and when I peek in at the window I see no bees. The colony was small and kind of limping along this summer so I am not surprised, but it is still disappointing. Now, I need to know if I should harvest honey that is in the hive. Can I leave the brood comb for the next colony to reuse? Do I bring it inside so wax moths don't destroy everything or just close it all up? Maybe I should leave it be and see what is left in the spring? I will wait to do anything until the next nice day when I can open it up and insure that the bees aren't just clustered in middle where I can't see them.

Robin answers:

I would like to tell you how I locate my bees when I can't open them: I use a doctor's stethoscope.

I've only used it twice, once to confirm my suspicions that the bees where gone in one hive; and most recently to happily hear the cluster in the hive I have left. I live in Wisconsin and it's way too cold to open the hive; hearing them in there sure relieved my anxiety.

Don writes:

Sorry to hear your bees may be dead. There is a lot of that going around this week. I talked briefly with Mo. She thinks she has lost a hive. David W wrote to me he's not seeing any activity in front for one hive and very little for the other. Today, I confirmed that the Osceola colony is dead. Anyway, please don't feel alone and certainly don't feel like you have failed in any way. I've been saying all year, some of these bees will die and that is eliminating genes that do not work well here in our geography and here in our time. We will all celebrate those that make it. The toughest time of year is just before the spring weather brings abundant food.

Illustration 219: No-cook candy experiment. Photo credit: Linda Chumbley

All the answers below assume the colony died of natural causes. Until we actually examine the dead-out, we do not know the cause of death.

Waiting for a warm day to verify they have perished is the correct thing to do. Maybe even give them several warm days of no activity before breaking in. I confirmed no activity 5 times before popping the top on Osceola today.

You could just leave it and see what's there come spring. Although that would not be my recommendation, it is a valid choice. I have no problems with letting nature run its course.

If they are dead, I would recommend you pull the honey, otherwise, it will likely be robbed. You have choices about how to use that honey. Human consumption is fine. I recommend, because you have another hive, that you find a way to store it safely (no crushing) in case you need it, someday, to feed to the other hive. Note: uncapped nectar is not honey and it will ferment and that can kill honeybees if they eat it. Leave the brood comb for the next colony to reuse. This is very valuable to you and your next colony. Treat it accordingly. If you are worried about wax moths, you can place the brood comb into a container inside. Wax moths are not a problem this time of year.

Linda writes:

I mixed up a batch of no-cook bee candy and buried a pollen patty in the middle. The picture is my Warré candy frame. I lined the frame with 1/2" hardware fabric to hold the candy. I used my stand mixer to mix 6lbs of sugar with a little more than a cup of water and a splash of vinegar. Way easier than trying to mix by hand.

Puttin' On the Candy

Saturday, 12/22/12, looks like a real possibility. Weather may cooperate. This is a mid-day event that moves quickly from apiary to apiary. I have to hit all these, south to north:

Whitehead's Warré Newport

Rinehart's Warré Wiley

Brown's Warré Quay

Crescent Lang Sugarloaf

Rock Creek Pumpkin 1.1

Hearteye Lang Violet

Willow Warré Franklin

Originally, I had planned Oak & Osceola. Oak we decided to experiment ... no candy. Osceola is already dead.

That's the full 50 miles from my most southern apiary to my most northern apiary.

Students should plan to participate at one site at least. You will learn this:

- candy installation procedure
- what does it look like when you put it on
- so you can compare what it looks like when you take it off
- what do bees look like at solstice
- compare this with spring also

Candy is going on today, 12/22/12. High will be 52 degrees F and winds are calm. I must move on this window of opportunity.

I'll start at David's house in Lakewood and work my way north. This must be done during the warmest part of the day today. Cannot start before 11AM and must be done before 3PM.

12/23/2012

Yesterday resulted in mostly good news.

I did not see every colony, but all those I saw were alive with the exception of Pumpkin 2, which MAY be dead, but I will have to check back. As you will recall, this is a 2nd year colony that entered winter with an assumed heavy mite load. We chose *not* to split in July, knowing that was a risk, but we were not excited by their performance, that is, no honey for harvest. If they are dead, it's sad, but not a big surprise.

I left candy for David to install on Newport and QE II.

I installed candy on Wiley, Quay, Cherokee (heavy population flying), Pumpkin 1.1, Sugarloaf and Violet, in that order.

Confirmed living Violet 2.

Unable to stop at Cottonwood Farm.

Overall, I'm very happy with only 2 colonies dead at solstice. The next eight weeks get tougher as we progress. Food stores only decline. There will be no further peeking until mid March. Now we wait.

Beekeeping Industry Transition

© steve kennedy 2014

We happen to be living at a time when a dramatic change is occurring in the beekeeping industry. The decade I'll call the twenty-teens will be known as an era boundary for beekeepers. The transition is from a few beekeepers keeping thousands of hives to thousands of beekeepers keeping a few hives. In the past, a few beekeepers, keeping sometimes tens of thousands of colonies, dominated the industry. Indeed, some of these persist, even today. However, the future promises to look much different. Today, thousands of hobbyist beekeepers are actively nursing a newly budding interest in keeping a hive or two of honeybees. Much like the transition that must happen for successful world economics, the future of pollination is, necessarily, small and local.

Colorado State Beekeepers Association (CSBA) is the umbrella organization nurturing many regional associations throughout our state. CSBA has set out to represent the current state of beekeeping both in terms of services provided and in terms of membership involved. This is quite challenging because, as you can imagine, the needs of a commercial beekeeper with 10,000 hives is quite different from the needs of a hobbyist with two.

Commercial beekeepers are concerned with pollination contracts in a widely dispersed geographic area. They worry about pesticide

Illustration 220: Rock Creek apiary sits in the middle of hundreds of acres of pumpkins. Photo credit: Don Studinski

use on large monoculture crops. They feed and medically treat very large numbers of colonies during tight time frames based on needs affected by climate and geography in addition to diseases, pests and parasites. They must cope with unique honeybee stresses related to long distance travel which restricts access to food and water.

Hobbyists are concerned with individual colonies, typically co-located in a single apiary, like a backyard. They are less concerned with the pollination provided and more concerned with the pollen being found and brought into the hive. They worry about pesticide use in relatively small urban or suburban areas. However, because the geography is typically densely populated, they are not able to verify pesticide usage among every neighbor within foraging distance. Many hobbyists focus on natural or organic beekeeping and avoid treating honeybees with medicine. At the same time, they may be more likely to feed sugar syrup or candy because they have the luxury of time to pamper their bees almost like pets.

Of course, there is a third category, the side-liner, who has substantial honeybee inventories, significantly beyond most hobbyists, but significantly less than most commercial beekeepers.

The point here is that the needs and concerns of various beekeepers are radically different and, consequently, the organizations that support those folks, like CSBA, are stretched thin to meet those needs. Similarly, as we go through the transition from big and geographically dispersed to small and local, the vendors that meet beekeeper needs, for example Denver Urban Homesteading, will be stretched, sometimes dramatically, to modify their services to meet changing needs. All beekeepers are a part of that changing landscape. You can contribute to everyone's success by patiently helping vendors to understand your needs.

Illustration 221: Midori's hive sits on the south side of her suburban home. Photo credit: Midori Krieger

How dramatic is the change? Let's consider some numbers. From 2010 to 2012, the CSBA membership has grown by more than 1400% in terms of human members. In 2010, members consisted of approximately 60, mostly middle-aged males[74], about 70% were commercial beekeepers (more than 150 hives).[75] Total hive counts are difficult to come by for our state, but in April of 2007, with less than 10 commercial beekeepers in Colorado, the Rocky Mountain News wrote, "Colorado, with about 28,000 colonies, ranks 20th in the nation for honey-producing colonies."[76] In 2012, CSBA members numbered over 850 humans of widely diverse age and both sexes are well represented. I could not find a source for a recent count on the number of managed hives, however, my guess is that the number of honeybee colonies in Colorado has not changed significantly. However, who is managing them and where has changed a great deal. This provides us with a small example of a trend happening throughout the USA.

74   ref: personal correspondence with CSBA
75   http://www.denverpost.com/news/ci_18284172, visited 12/23/2012
76   http://m.rockymountainnews.com/news/2007/Apr/30/bee-losses-a-mystery/, visited 12/23/2012

There's another significant factor at play. Feral honeybees are in serious decline. An estimated 70% of feral honeybee colonies have now died.[77] Some locations indicate that the feral honeybee population has completely disappeared.[78] The USDA CCD Steering Committee said in 2007, "parasitic mites have destroyed most of the feral honeybees across the United States."[79]

What does this mean? It means that a substantial portion of the pollination services required in Colorado are now provided by hobby beekeepers. It means that were it not for those hobbyists, a substantial portion of the fruit and vegetable crops produced for family and neighborhood consumption would be missing. Feral honeybees are gone and commercial beekeepers do not engage at the neighborhood level. It means you, the hobbyist, are an important piece of a much bigger and very important system, a complex ecosystem that is struggling to hold on to a fragile balance. It means your work is important, even though no one is likely to thank you. This is true way beyond Colorado. This is true throughout the World.

One particularly good news item is that, due to this growth in beekeeping interest, it is much more likely you can find someone close by that shares your interest, perhaps passionately. You can help and learn from each other. You need not feel alone.

There is much to be gained by joining your local beekeeping club as well as your state and national associations. Guest speakers, access to affordable training and new friendships with your fellow beekeepers are some of the obvious benefits. Perhaps less obvious is the opportunity to turn off that television and read some of the excellent articles peers are writing for our journals.

Remember, beekeeping is beneficial to your health. In this case I'm ignoring the medicinal value of the hive products. I'm just talking about the stress reduction and psychological benefits of reconnecting with nature. The bees provide you a compelling excuse to get in touch with the seasons and to think about natural cycles as well as acute natural events on a regular basis. Many beekeepers notice an improvement in their ability to achieve a calm state. It may be that the bees have a calming effect on humans, or it may be because we quickly learn the benefit of approaching the hive in a calm state and therefore are practicing our own self-calming abilities. Whatever the case, becoming a beekeeper may be one of the best things you've ever done for your health. What else in your life is actually encouraging you to slow down rather than further stressing you out?

---

77 http://www.apexbeecompany.com/honey-bee-facts/, visited 8/3/2014
78 http://www.bees-and-beekeeping.com/honey-bee-deaths.html, visited 12/23/2012
79 http://www.fs.fed.us/r6/invasiveplant-eis/Region-6-Inv-Plant-Toolbox/2008-New/Colony-Collapse-Disorder.doc, visited 12/23/2012

Epilogue

© steve kennedy 2014

Writing a book is a big project. I have an entirely new appreciation and respect for those who have preceded me in this endeavor. Today is March 3, 2014, which means I'm now into my third year of working on this book. I still have a number of formatting issues to overcome and I have not yet found a publisher, so it may be some time before this becomes available for purchase. Learning to use the software to format things as I wish turns out to be as much of a challenge as writing, finding references and getting the photos.

Obviously, much has happened with respect to the colonies living in 2012. For those curious about specific results and specific colonies, I'll close with a status report. Although your results will vary, I can promise you that if you stick with beekeeping for several years, you will experience both successes and failures, survivors and dead colonies. My wish is that you find joy in it all and that this text has added to that joy in some way.

Illustration 222: Hive build of 2014 where we built a modified Langstroth design using an upper entrance. We built 15 hives in one day. Photo credit: David Whitehead

Mama Birch was one of my best for 2011, 2012 and 2013. She produced a number of daughters, most were successful, some I even sold to other beekeepers. She finally died due to a robbing event in the fall of 2013.

Birch 2 and Birch 3 were both 2012 daughters of Mama Birch. They produced a little honey in 2012 as first year colonies, which is somewhat unusual. They both survived the winter of 2012/2013 very robustly. By April of 2013 they were roaring with 50,000 bees each.

Alas, the weather turned out to be brutal that month and single digit freezes killed all the blossoms. I hesitated to feed them. I kept thinking "The blossoms will come any day now." That turned out to be a big mistake. They both died of starvation with so many mouths to feed.

Superior was a great colony 2010, 2011 and 2012. They made it to spring of 2013 and I wanted daughter queens. Unfortunately, I tried to feed them with a feeder that ended up leaking at the front door. This is basically a guaranteed recipe for robbing. Nature never ignores an opportunity like that and they died trying to defend their stores.

Queen Elizabeth II and Newport were both on the same property. QE II was a daughter queen that never really thrived. Newport was a swarm that did fine for a first year colony, but did not produce harvest. They both died in the fall of 2012 from symptoms that looked like colony collapse disorder (CCD). Editorial note: I don't believe CCD exists. It's just a set of symptoms consistent with systemic poisoning. Many beekeepers are quick to blame mites. We didn't fine any mites in the remains.

Braden was a nuc from Texas. They produced a small harvest in 2012, but did not survive the winter.

Rinehart was also a nuc from Texas. They produced no harvest in 2012, but not only survived the winter, they also threw a swarm in 2013 and are still alive today. That means the old queen flew with the swarm and a daughter is there now.

Wiley was a swarm in 2012 that failed to make it through their first winter.

Sable, Osceola and Oak were all 2012 swarms that never thrived. They failed to make adequate comb, failed to grow a good workforce, failed to bring stores adequate for winter and, to no surprise, they died of cold or starvation or both.

Quay, a 2012 swarm, didn't produce harvest in 2012, but was strong in spring of 2013 and threw a swarm. They are still strong today and I expect to get a spring harvest from them someday soon.

Quay 2 never really performed and we merged them with Cherokee, also too weak to winter. We called the merged colony Cherokee to eliminate the two Quay colony confusion. In 2013, Cherokeee produced two strong daughters. One was sold, but died fall of 2013. The other daughter is thriving as is the mother. Mom will be one of my honey producers in 2014 and I'm hoping for granddaughters.

Cottonwood died of robbing in the fall of 2012. This one was not my fault, but nature's choice.

Pumpkin 1, Pumpkin 2, Arapahoe and Pumpkin 1.1 were the unfortunate victims of an accident. The farmer of that land allowed his cows to get into the apiary where they knocked over the hives; some got crushed, and this was in the middle of December. I cleaned up the mess to the best of my ability and three of the four actually made it to spring, but then all died of cold or starvation or both.

Clarkson has been a real go getter. A 2012 swarm, they produced three daughters in 2013. One of those daughters was the first successful 2013 split in early April. That daughter is also thriving along with mom. The other two daughters have since died due to queen failure.

Violet and Violet 2, both 2012 swarms, died of queen failure early in 2013.

Franklin and Queen Street also both died of queen failure early in 2013.

I spent much of 2013 trying to recover from my losses. You will have years like that too. "Two steps forward, one step back" as the old saying goes. I'm approaching spring of 2014 with 10 strong colonies. I entered November's no peeking season with 16. That's about right in the average for natural beekeeping where we have been seeing 30% - 40% losses every year since 2006. I once compared my losses with those using chemicals to control mites and disease. They are also loosing about 30% every year. For me, that potential 10% difference is not worth spreading more chemicals over this precious earth.

Whether you choose to do permaculture style beekeeping or you choose a more traditional route, I hope you enjoy your journey. You have made it to the end of this 2012 adventure. If you would like to join in as a student, the beekeeping adventure continues via my yahoo group which you can join for free here: https://groups.yahoo.com/neo/groups/BeekeepingStudents/info.

Glossary

© steve kennedy 2014

A

Abscond – the honeybees abandon their home. This may occur soon after installation, or it may be after they have lived in the hive for a long time. The queen and all bees leave. This is not CCD.

Adult – a honeybee having chewed through the cell cap and emerged. In the case of a worker, such a bee is immediately ready for work.

After swarm – see secondary swarm.

American foulbrood (AFB) – a deadly bacterial infection.

Apiary – the land where one or more managed honeybee hives are kept.

Apitherapy – the medicinal use of hive products.

B

Back-filled – workers have filled brood comb cells with honey and bee bread.

Ball – a verb describing workers killing a queen using heat. The workers surround the queen in a ball while vibrating their wing muscles creating kinetic heat.

Bearding – a crowd of bees hanging on the front of the hive to reduce the temperature inside the hive.

Bee bread – pollen mixed with enzymes and honey, slightly fermented for long-term preservation. This is the honeybee source for protein.

Bee brush – a soft bristled brush intended for moving bees when you need to. A large feather is an alternative.

Bee package – a small wooden container holding about three pounds of honeybees. The container is ventilated using a screen material so the bees can regulate their temperature. The container also includes a can which holds sugarwater so the bees can eat and drink.

Beek – short for beekeeper.

Beekeeper – a human that chooses to manage bee colonies.

Bees – a term used interchangeably with honeybees, but may also refer to many other types of bees.

Bee suit – protective clothing that prevents the beekeeper from getting stings.

Bottom board – the bottom of a hive.

Bridge comb – comb built by honeybees across two different frames forming a bridge from one frame to the next.

Brood – the new bees being raised in the nursery area of the comb. This area is referred to as the brood area, or just the brood. It includes eggs, larvae and pupae under caps.

Burr comb – comb built on box sides and other places where beekeepers don't really want it. It's usually not full comb, but a portion of the comb shape that resembles a burr.

C

Candy super – a super box just a few inches high giving enough room for brick(s) of candy inside with bee space around the candy, but too small for frames. The idea is to make as little added space, which must be kept warm, as possible.

Capped brood – cells containing pupae. Under the wax cap, a larva spins a cocoon and changes into an adult bee.

Capped drone – brood capped with a higher, puffy cap, sometimes called bullet shaped.

Capped honey – fully cured nectar, reduced to 18.6 or less moisture content, is honey. Bees put on a wax capping when it is done.

Capped queen – the peanut shaped cell in which a queen transforms from a larva to an adult.

Capped worker – brood capped flush with the top of the cell.

Chalkbrood – a fungal infection resulting in death to brood where the bodies take on a chalk like look and feel.

Checkerboard – intentionally introducing empty frames every second frame in an effort to eliminate the swarm impulse.

Cluster – a tight formation of the colony into a "ball" of sorts where they shiver to keep themselves warm.

Colony – a family of honeybees, mostly workers, one queen and about 5% - 15% drones during the warm season.

Colony collapse disorder (CCD) – the name given to a set of symptoms that showed up in 2006 killing many, many colonies suddenly.

Comb – the hexagon cells of wax built by honeybees for storing food and raising young.

Crop – the vessel which carries nectar back to the hive.

Crushed comb – a means of removing honey from comb that destroys the comb and uses gravity to perform the separation. The crushed wax is not wasted, it is generally used to make other products.

Cut comb – whole honeycomb filled with honey carefully cut to a container size. Humans eat it whole. It generally demands a premium price because the comb is valuable to the honeybees and the beekeeper.

D

Dead-out – a hive containing a dead colony.

Deep – a box intended to hold frames, generally used for the brood area of the hive.

Diploid – an egg that has been fertilized by a sperm, therefore, containing the full set of chromosomes, both male and female.

Drawing out new comb – when honeybees create new comb, whether on foundation or natural comb, we refer to the activity as drawing out comb.

Drone congregation area (DCA) – an area about 50 feet high where drones of many colonies gather to await virgin queens. Somehow, both the drones and the queens know where to look for this place.

Drone frame – a frame with foundation pressed with drone sized cells. When the bees draw out comb on such a frame, all the cells are drone size. A queen will lay all drones on such a frame found in the brood nest area.

Drone layer – could be a laying worker or a queen that did not get adequately fertilized.

Drones – male honeybees. They have no stinger. They do no housework. They have only one purpose: to fertilize a virgin queen.

E

Egg – the first stage of development for a new honeybee. Say this aloud: An egg is an egg for three days. Repeat that. Memorize that.

Emerge – the act of chewing through and crawling out of the wax capping. The birth of a new honeybee. Adult bees emerge.

Empty cells – honeybee comb is made of wax which form cells. Cells are intended to hole things like larvae, bee bread and honey. When they are not currently holding something, we call them empty cells.

Escape board – an inner cover that will allow bees to pass one direction, but not return. This can be used to remove bees from harvest honey supers.

Extractor – a centrifuge tool for spinning honey frames which causes the honey to come out of the cells. This might be motor driven or hand cranked.

F

First bloom – the first fruit tree blossom you see open each year. Make a note of the date. This will help you with your planning in future years.

First inspection – officially ends no peeking season, approximately mid March. This is your first look at the colony during the calendar year. It calls for split assessment and hive cleanup.

Flow – a way of referring to the time of year when nectar is readily available in the blossoms. This is the time when the honeybees have access to abundance and can make honey which can be harvested. See nectar.

Foundress – a varroa mite in a honeybee larva cell where she will lay eggs.

Foundation – wax or plastic pressed with the honey comb shape intended to be placed in a frame on which the honeybees will draw out the comb.

Frame – the wooden structure on which the bees build their comb. It may include foundation or not. Sometimes, top bars are referred to as frames because of their similar purpose.

Fuel – flammable material that can produce cool white non-toxic smoke.

Fume board – a felt lined inner cover which allows stinky chemicals to be applied which drive the bees away. This can be used to get bees out of harvest honey supers.

G

Guard – worker performing the task of defending a hive opening.

Gated – some food grade buckets have had a gate installed which allows honey to flow out in small amounts. This tool helps with filling jars. We refer to a bucket with a gate as a gated bucket.

H

Haploid – an unfertilized egg, therefore, containing only half the chromosomes, just the female chromosomes.

Hatch – eggs hatch. It means break open to allow what is inside to expand. In our case, what's inside is a honeybee larva.

Helmet – the hard hat part of a hat and veil combination.

Hemolymph – bee blood.

Hive – the home of honeybees. What lives inside is a colony.

Hive mind – the collective decision of the colony about what to do and when. Decisions are always cooperative and the well being of the colony always takes precedence over well being of the individual. Honeybees do not know selfishness.

Hive tool – a handtool used to pry apart boxes and frames that are stuck together with propolis and wax.

Holometabolism – honeybee metamorphosis. Complete transformation from egg to adult.

Honey bound – a condition where the colony is short on space to lay new eggs. The honey is taking up all the space.

Honey stomach – a slang term for the honeybee crop.

Hood – a plastic reinforced veil usually attached to a jacket or body suit.

Housekeepers – workers performing the the task of cleaning and building comb.

I

Increase – using the split process to multiply your live stock.

Inner cover – rests over the top box in a stack. Provides some insulation and generally includes a hole useful for feeding.

L

Langstroth – the hive type typically thought of as "traditional" which was invented by Lorenzo Langstroth in 1851. It uses removable frames and encourages the bees to build vertically.

Larva – the worm stage of honeybee development. Days 4 through 9 (for female) or 11 (for male). The plural of larva is larvae.

Laying worker – a worker that chooses to lay eggs. Laying workers are always present. Generally, other workers are cleaning up the mess laying workers make. In the absence of a queen, the laying worker becomes noticeable to the beekeeper because the mess is not cleaned up.

Location – the place (apiary) you will be keeping your honeybees. Key factors when choosing a hive location include wind, sun and water.

M

Merge – the act of combining two or more honeybee colonies into one.

Metamorphosis – the complete transformation of a new honeybee from egg to adult.

Micro climate – a relatively small area with unique weather. This can be as small as a few square inches.

Mouse guard – a tool which blocks the hive entrance from access by mice. This tool must allow honeybees to pass, but block a mouse from passing. Although a mouse can pass through a flat space that is only ⅜ inches deep and a few inches wide, it doesn't seem to be able to get through a ½ inch square.

N

Nectar – the sugar syrup produced by plants which honeybees collect and bring to their hive where they turn it into honey. Nectar is 80% water. Honey is 18.6% or less water. See flow.

Neonicotinoid – a type of poison. This is one type of systemic poison, there are many.

No peeking – a time when it's important that the beekeeper not break the propolis seal between hive pieces. Leave the bees alone. Breaking the seal causes them more work to put it back. This begins November 1 and ends mid March. The cold months of the year.

Nosema – a fungal bee gut disease which causes the bees to have diarrhea. There are multiple forms of nosema. Generally, bees can recover, but it can be deadly.

Nuc (short for nucleus) – commonly used as a short way of saying nucleus which refers to a small, two to six frame, honeybee colony. Typically, nucs are involved with the split process. Beekeepers might purchase bees in a nuc. In the beekeeping world, nuc can refer to the little box the colony is living in, but nuc can ALSO refer to the small colony of bees. This can be confusing, so try to be aware of the context when the word nuc is used.

Nuptial flight – the mating flight of a sexually mature insect. For honeybee queens there may be several flights.

Nurse – a worker bee performing the task of feeding larvae and warming the brood.

O

Orientation – for bees, this is honing in on the hive entrance. Foragers memorize where the entrance is so that the last part of their flight home is instinctual. For hives, this is how the hive is positioned, for example, with the entrance toward south.

Osmotic effect – osmosis establishes or maintains the stable state of equilibrium.

P

Package – a short way of referring to a package of honeybees. See bee package.

Permaculture – a way of living that strives for harmony with nature.

Primary swarm – the first swarm of a season which will contain the over-wintered queen.

Propolis – very sticky sometimes the consistency of hard candy, this is antibiotic, antimicrobial, anti-fungal and antiviral tree resin gathered by bees to surround themselves and seal cracks.

Propolis envelope – the coat of propolis laid down on all interior hive walls as a environmental defense against bacteria, fungi, viruses and microbes.

Proventriculus valve – prevents or allows nectar to pass into the honeybee ventriculus (stomach).

Pupa – the last transition before emerging as an adult. Pupa develops inside the cocoon which is under the wax cap on the cell. The transition is from worm, to fully developed honeybee body. The plural of pupa is pupae.

Q

Queen – the one female in a colony fully capable of reproduction. Having mated with 10-20 drones in the first or second week of life, she lays eggs for the reminder of her life without further mating.

Queen cell – a large peanut shaped cell made specifically to hold the large larva and pupa of a queen.

Queen cup – an invitation to lay an egg destined to be a new queen.

Queen excluder – installed between a brood area (below) and a honey area (above) this device prevents the queen from laying eggs in a honey area intended for harvest.

Queen right – a hive containing a honeybee colony with a healthy, laying queen is said to be queen right.

Quilt – the box intended to hold insulation material which sits near the top of a Warré hive. Only the roof goes over the quilt.

R

Retinue – the workers attending to the queen's needs: feeding, directing her laying and removing her waste.

Robbed – honeybees must be constantly vigilant about defending their food stores. Nature never lets anything go to waste. Any undefended food source will be utilized by some opportunist. Robbed honeybees may or may not survive; it can be serious.

Robbers – honeybees sometimes rob other honeybees. If they are able to penetrate a competing hive defense, then the food is easily stolen and the robbed colony may or may not survive. Members of different colonies fight during such an event.

Royal jelly – the substance nurse bees feed to larvae during a variable portion of the larva stage depending on the larva being a queen, a worker or a drone.

S

Secondary swarm – swarm with a virgin queen; all swarms after the primary swarm. Also know as a virgin swarm or an after swarm.

Smoker – a tool used to contain a small, smoldering fire which a beekeeper can easily carry as hives are inspected. The smoke has a calming effect on the bees because it interferes with their ability to communicate, for example, the "defend" message.

Solar melter – a wooden box lined with tin and covered with a window. It's intended for melting old or broken comb. Wax drips out the bottom into a container.

Split – the process of starting with a single colony of honeybees and creating multiple colonies.

Stinger – the weapon of a honeybee.

Strainer – a wire mesh tool that helps filter out bits of wax and pollen. We still call this honey unfiltered or raw because it has not been heated and pushed through a paper filter. Strained is not the same as filtered.

Sun – the second thing to consider when choosing a hive location. Bees need full sun in winter for sure. Sometimes it's nice to have a deciduous tree on the south side so the bees get a break from the heat in summer.

Super – a box over, superior to, the brood area. Generally used to refer to boxes intended for honey harvest.

Superorganism – a single living entity made up of many living entities.

Supersedure – refers to the process honeybees execute when they choose to replace their queen.

Swarm – the birth of a new colony. They must find a home. While homeless, they are in their most docile state. With no home to defend, their instinct to sting is generally not triggered. (swarming, swarmed)

T

Telescoping cover – sits on and around an inner cover and the top box of the stack. Helps prevent wind from blowing off the top.

Top-bar hive (TBH) – a hive, generally inexpensive to build, which uses a horizontal orientation rather than vertical and which employs removable wooden bars across the top on which the bees build comb. This hive type seems especially popular with natural and hobby beekeepers.

Topology – the terrain or landscape. Within a hive, it's how the frames are arranged and how the comb is built.

V

Varroa destructor – a parasitic mite.

Veil – a head covering garment made of netting which keeps the bees out of your face.

Venom – the poison a honeybee injects into a victim when she decides to use her stinger. Poison seems too strong a word in this case because that venom also has positive medicinal value. But, it can kill under certain circumstances.

Ventriculus – honeybee stomach where food nourishes the honeybee.

Virgin swarm – see secondary swarm.

W

Warré – the people's hive invented by Emile Warré. Similar to Langstroth in that it is oriented vertically with boxes that stack. Similar to TBH in that the bees build natural comb without foundation.

Water – the third important factor you must consider when choosing a hive location. Bees will find close convenient water. If that turns out to be your neighbor's faucet, then they may not be happy with you. Just think about where your bees will get water. It must be relatively close, not more than a mile.

Wet honey – capped honey where the honey is actually touching the cap. Just "honey" or just "capped honey" has a bright white wax cap. "Wet honey" has a dark cap because the honey is touching.

Wind – the most important factor when choosing a hive location. Seek a wind break on the sides from which the prevailing winds come.

Workers – female honeybees not capable of reproduction. These bees do, literally, all the work. They are smaller than males, drones. They are also smaller than the queen. They can sting.

Alphabetical Index

About the Author
Donald P. Studinski

Don Studinski, dba Honeybee Keep, is a permaculture enthusiast and member of the board of directors at Living Systems Institute (LSI) where he applies permaculture philosophy to beekeeping. Apiaries under Don's management are located from Golden to Erie, spanning about 50 miles. Honeybee Keep manages Colorado's first Certified Naturally Grown apiaries. Don's beekeeping articles have been published in Bee Culture magazine and on-line at Honeybee Haven, Peak Prosperity, Bee Informed Partnership and Selene River Press.

As a beekeeping mentor, Don provides advice and counsel for students throughout the United States. You can reach him using dstudin@yahoo.com. Colorado Bees for Colorado Beekeepers is Don's "produce local bees" project which provides nucleus colonies for sale. Learn more about beekeeping, read free articles and see all the products and services provided by Honeybee Keep at HoneybeeKeep.com.

Don has a BS in Computer Science and an MS in Computer Information Systems. He spent his computer career working for IBM, StorageTek, and McKesson as a programmer, manager and director. He has also owned and operated his own t-shirt and embroidery business. Today, Don spends his time as a beekeeper, mentor and community building activist. He removes bees and wasps from structures, collects swarms, sells honey, performs public speaking, provides honey extraction and provides beekeeping consulting.

You can join Don's Yahoo Group. It's free. Read lessons and ask questions. Share experiences with other students.

https://groups.yahoo.com/neo/groups/BeekeepingStudents/info

To Contact the Author, Donald P. Studinski:
by email: dstudin@yahoo.com
by postal service: POB 1995, Broomfield, CO 80038
author's website: www.HoneybeeKeep.com

CPSIA information can be obtained
at www.ICGtesting.com
Printed in the USA
BVOW05s1210180117
473800BV00016B/145/P